2c 5/97

Plastic

Plastic

THE MAKING OF A
SYNTHETIC CENTURY

Stephen
Fenichell

HarperBusiness
A Division of HarperCollins*Publishers*

HarperCollins books may be purchased for educational, business, or sales promotional use. For information please write: Special Markets Department, HarperCollins Publishers, Inc., 10 East 53rd Street, New York, NY 10022.

Designed by Alma Hochhauser Orenstein

Library of Congress Cataloging-in-Publication Data

Fenichell, Stephen.
    Plastic: the making of a synthetic century / Stephen Fenichell.
      p.   cm.
    Includes index.
    ISBN 0-88730-732-9
    1. Plastics—History.   I. Title
    TP1116.F46   1996
    668.4—dc20   96-364

96 97 98 99 00 ❖/HC 10 9 8 7 6 5 4 3 2 1

For
CAROL
and
LOISA

"I just want to say one word to you, Ben. Just one word."

THE GRADUATE (1967)

Contents

Acknowledgments

To my publisher, Jack McKeown, and my editor, Adrian Zackheim, for their unstinting support of this book from its inception to (and beyond) its completion; to Dorothea Herrey of International Creative Management for "getting it" from the beginning and maintaining an unflagging interest in its realization; to Suzanne Oaks of HarperCollins for her careful reading and insightful, incisive suggestions; to Janet Dery of HarperCollins for keeping innumerable loose ends knotted and tied; and to my beloved in-house editor, Carol Goodstein, who kept a close eye on the details while never losing sight of the big picture.

I'd also like to thank the dedicated librarians at the Science & Technology Room at the New York Public Library for putting up with a year of tough requests, as well as the librarians who assisted with fulfilling research requests at the libraries of The Stevens Institute in Hoboken, N.J., Brooklyn Polytechnic, and The Hagley Museum and Library in Wilmington, Delaware.

To Helen S. Carothers, widow of Wallace Hume Carothers, for permission to publish extracts from W. H. Carothers's letters to John Johnson, in the manuscripts and archives department of The Hagley Museum and Library. To those employees of DuPont, Ford, General Motors, Dow Chemical, Hoechst Celanese, Rohm & Haas, Kodak, and Mattel for their aid in compiling illustrations. And to the photographer Fred Kligman, for permission to use several of his superb plastic photographs.

Plasticland

Courtesy of Fred Kligman

S cientists said it could be never be seen by the naked eye, even in the night sky, some two hundred miles above the equator. But the luminous tendril was clearly visible to hundreds of watchers, about an hour after sunset and an hour before sunrise, in the skies above Florida, Texas, New Mexico, and southern California. As the space tether whirled eastward along the equatorial belt, passing Hawaii en route to Hong Kong, India, Pakistan, Saudi Arabia, Puerto Rico, and Baja California, it showed up against a field of stars and pinpoint planets as a bright shining line, roughly three lunar diameters long.

"This is the first man-made object in space visible as something other than a point of light," exclaimed Dr. Joseph A. Carroll—the scientist who first proposed the space tether experiment to NASA a decade before. "I couldn't see the space shuttle, but I could see fifteen pounds of string."

Those fifteen pounds of string had been spun from a high-density, high-strength polyethylene called Spectra into a fiber twelve miles long but just one thirty-second of an inch thick. As a polymer whose molecules are all oriented in the same direction, Spectra enjoys one of the highest strength-to-weight ratios of any material on earth.

In the near future, space tethers as strong or stronger, as thin or thinner, will replace cumbersome, costly rockets for much of the heavy lifting in space.

Polyethylene, the plastic used to make space tethers, is also used to make grocery and dry-cleaning bags. Discovered by British scientists during World War II, it helped the Allies win the air war over the Axis powers in Europe. Without polyethylene for lightweight radar insulation, the RAF would have been unable to install radar on its airplanes. Without radar, the British would never have enjoyed the critical edge in surveillance that helped them overcome the Nazis' overwhelming superiority in aircraft and manpower.

"I just want to say one word to you, Ben. . . . *Plastics*. There's a great future in plastics. Think about it."

"Yes, sir," Benjamin blankly replied. But I very much doubt he did. What would a bright, sensitive guy like Benjamin think about plastic, except to dismiss it out of hand as an alien substance foisted upon us by unscrupulous manufacturers with an eye toward cutting corners? If not strictly emblematic of all that is fake and wrong with the modern material world, at least an uninvited guest at the party.

Most of us, after all, are materialistic enough to love certain things. Things, by and large, made from materials that like us chip, wear, abrade, erode, distress, and die. Whether they burnish, tarnish, rust, mellow or patina, soften or fade, most natural things grow graceful with age.

But not plastic. When it wears out, it cracks. That's it. You chuck it away without the slightest sentimental attachment, except, perhaps, the fleeting hope that it will go away. But it won't go away. Instead, it sticks around, like a pest at a picnic. The Rasputin of modern materials, you can break it, chop it, dice it, shred it, burn it, and bury it, but it stubbornly refuses to die.

As fashion arbiter Tatiana Lieberman explained to her daughter, the author Francine du Plessix Gray, when she wondered why her taste-obsessed mother had decorated their living room with a suite of cheap plastic garden furniture—"Plastic *ees* forever." Like it or not, plastic *is* permanent. The question becomes: How do you handle it?

An official of the Office of Naval Research recently remarked: "Petroleum-derived plastics were *designed* not to biodegrade. They survive in landfills forever, just like they were supposed to." An advisory notice issued to campers in New York State parks, posting the average decomposition times for "typical camping litter," confirms this happy hypothesis.

Leather shoe: 5–40 years
Plastic bag: 10–12 years
Nylon cloth: 30–40 years
Plastic container: 50–80 years
Plastic foam: *never*

In 1989, the absolute nature of that *never* prompted Tanya Vogt, a fifteen-year-old high school student from West Milford, New Jersey, to launch a nationwide campaign to ban polystyrene foam cups and clamshell containers from the high-school cafeterias of New Jersey. The movement rapidly gained steam. The Berkeley, California, City Council predictably followed suit with its own ban. Before the year was out, McDonald's, which had vigorously defended its right to keep burgers warm in any container it pleased, caved. From coast to coast, Big Macs hit America's Formica counters wrapped in recycled brown paper bags.

Still, plastic's legendary endurance has provided us with a few unforeseen benefits. In January 1992, a freak ocean squall washed a container of cargo off a freighter crossing the international date line. The container broke up in mid ocean, releasing twenty-nine thousand plastic bathtub toys being shipped from Hong Kong to Tacoma, Washington. Months later, hundreds of blue turtles, red beavers, yellow ducks, and green frogs were sighted washing up on the shores of Sitka, Alaska. For the next year, thousands more washed up along a five-hundred-mile stretch of the Gulf of Alaska coast, giving oceanographers "the greatest boon for research on North Pacific patterns and currents since 61,000 Nike athletic shoes had been spilled in the same area two years before," according to Reuters. What became known to the marine research community as "the quack heard round the world" enabled scientists to adjust their computer models of the northern Pacific tides to account for the effects of the wind.

Seven years after the sunken hulk of the *Titanic* was discovered by a French-American salvage expedition, the French government gave would-be claimants of lost property—heirs of the passengers— three months to provide proof of ownership. The treasure trove raised from the *Titanic* featured gold and silver wrist- and pocket watches, buttons, bracelets, jeweled necklaces, rings, tiepins and hairpins, gold pince-nez spectacles, leather goods, several hundred English coins, ivory combs, mirror cases, and hairbrushes—all fashionable accoutrements of upper-class life circa 1912 in England and America.

But what most impressed Charles Josselin, secretary of the French Merchant Marine, was that out of that long list of everyday objects, not one thing was made out of plastic. "That, if nothing else," he quipped at a press conference, "shows how much times have changed."

Times *have* changed. In 1816, Christian Thomsen, a Danish

museum curator, coined the terms Stone, Iron, Bronze, Copper, and Steel Ages to illustrate his insight that human development is inexorably linked to the control and manufacture of materials. That if we are not what we eat, we are what we make. The Industrial Revolution ushered in the ages of Steam and Steel.

Our postindustrial epoch frequently goes by the handle Information Age. But it could just as easily be called the Plastic Age. Plastic provides us with the material prerequisite for information storage and retrieval, both analog and digital. From photographic film to audio- and videotape, from computer discs to CD-ROMs and CDs, plastic not only imitates natural materials, it allows us to recreate an entirely new world of the visual and aural imagination and record it for instant replay, as well as for posterity.

In 1979, the global volume of plastics production outstripped that of steel. At precisely that point in our industrial development, we entered the Plastic Age. In the five decades since the end of World War II, plastic has crept unceasingly, and often invisibly, into our homes, cars, offices, even our bodies. Some of us have plastic hearts, joints, valves, limbs.

Plastic has become the defining medium of our Synthetic Century precisely because it combines the ultimate twentieth-century characteristics—artificiality, disposability, and synthesis—all rolled into one. The ultimate triumph of plastic has been the victory of package over product, of style over substance, of surface over essence.

Plastic owes its bad rap and its bad rep to the undeniable fact that so many of its most fundamental applications have been so chintzy and sleazy. From plastic pocket protectors to AstroTurf, the more miraculous aspects of plastic have been buried beneath a superficial veneer of glam and gloss. What might have been wondrous, rich, and strange, the "essence of alchemy" in Roland Barthes's phrase, has never lived down its humble origins as a cheap substitute.

A few Continental contrarians do profess to admire the stuff. For its tenacity, its versatility, even its sublime vulgarity. Plastic being, of course, the quintessence of Pop. Madonna likes plastic— she wrapped her book *Sex* in it. Plastic is very vernacular. By making synthetics cheaper by the dozen, it's turned nature into a luxury. The Italian philosopher Ezio Manzini lauds plastic's infinite capacity to reinvent itself, to mimic natural materials, as "the essence of its authenticity." French cultural critic Roland Barthes praises plastic's "quick-change artistry"—its chameleon-like ability to be "buckets as well as jewels."

But in America, where plastic has made the greatest cultural inroads, high-IQ types make despising it a point of pride. The very word has become a term of derision, signifying anything in contemporary life that is postmodern and grotesque. Historian and biographer David McCullough, opposing the Disney company's now abandoned Civil War theme park, maintained that to replace "what little we have left that's authentic and real with *plastic history* would be sacrilege."

"It's plastic," muttered New York City councilwoman Kathryn Freed, dissing the shopping mall at Southstreet Seaport in lower Manhattan, which failed in its promise to properly harbor a museum of maritime life. Columnist William Safire calls Singapore a "plastic utopia," meaning that it isn't a real one. Visitors to eastern Europe are advised to hurry up and come before the region gets "plasticized." Airline food is routinely described—with good reason—by one word, and one word only: plastic. Processed cheese foods, we suspect, may *be* plastic. Frankly, we'd rather not know.

Personally, I've gone back and forth. Last year, in a fleeting fit of antiplastic passion, I rushed out to buy a pure cotton shower curtain, a varnished oak toilet seat, a wood-handled toothbrush with genuine bristles—the kind they pull out of the backs of pigs in Third World countries. But I soon regretted my reckless embrace of Mother Nature. After my next shower, the new curtain wouldn't dry. I had to buy a nylon liner to protect it from water. My beautiful bristle toothbrush also refused to dry. The idea of inserting moist animal parts in my mouth began to seem a little too *au naturel*. The final indignity occurred when my oak toilet seat cracked, setting me back another forty bucks. A full-fledged plastic purge is a luxury few of us can afford.

Because plastic is so cheap, it's become a white trash connoisseur's enduring delight. Think Ban-Lon, Orlon, Corfam, Lycra. Leatherette. Pink flamingos. Kitchen cabinets made of "polyester panels enhanced with natural foliage and butterfly embedments." Plastic Statues of Liberty in Woolworth bins. Plastic pietàs. "I don't care if it rains or freezes, long as I got my plastic Jesus right up on the dashboard of my car." Polyester leisure suits. Stickup Dixie-cup dispensers.

A pitched Vinyl War still rages in the pop music industry over the sonic qualities of vinyl versus the arrogant sterility of the CD. For grunge bands Nirvana and Pearl Jam, it's all a question of authenticity. And though it may come as something of a surprise to David McCullough, vinyl has in recent years (to some ears) attained cult status as the last repository of all that's real and funky about rock. Because vinyl degrades, it's intrinsically more likable than the virtually scratchproof polycarbonate CD, because it inadvertently immortalizes sonic imperfections, snap, crackle, hiss, pops, and all.

Which may help to explain the new Chicago club called "Vinyl." The seventies retro-chic Greenwich Village bar "Poly Esther's." The new Soho boutique, "Smylonylon," dedicated to offering sixties and seventies synthetic clothing—vintage but virgin—preferably made of nylon. "Nylon has a very big sexual connotation going on with it," insists owner Chris Brick. "There's something about the fabric that makes the girl or woman soft."

From Bakelite bracelets to plastic purses from the fifties, the Synthetic Century has in recent years begun to spawn its own breed of nostalgia. Fashion designer Betsey Johnson waxes wistful over her plastic dresses, quintessential sixties:

> The Kit Dress was a cellophane slip with metallic stick-on shapes you could decorate it with. My Growing Herb Dress was made of something like Handi Wipe Fabric with little seeds embedded in it. When you watered it, it actually sprouted and it became this fuzzy little green thing.

At ten, I learned to love the rugged Bakelite barrels on Parker ballpoint pens. They came in aqua and olive green and seemed and felt indestructible—which I imagine they were, under most student circumstances. I liked clutching and running my fingers along our basic black Bakelite rotary-dial phone. It had the healthy heft of a club and a tympanic sound to go with it. Pulling the plastic bits off the soles of my foot pajamas. Gazing awestruck at the leather grain of the padded blue dashboard of our Chevy Nova, wondering why it was padded, why in God's name the makers saw fit to give it fake leather grain.

So we do have fond feelings, some of us, about plastic. But plastic's tactile appeal is felt most strongly, I've come to believe, by the more aberrant among us—anal-retentives, for example, and qualified sado-masochists—though naturalist B&D freaks prefer real rubber.

Dr. Luc Jouret, the Belgian physician who in the year 1984—appropriately enough—founded the Order of the Solar Temple in Switzerland, took plastiphobia to the utmost extremes. The night before his cult's collective suicide a decade after its founding, Jouret was spied in the Swiss mountain village where he maintained his headquarters buying large quantities of plastic bags. The purchases were not regarded as unusual, since Jouret and other cult members had bought plastic bags in bulk many times.

Only after their collective apocalypse did the shocking truth emerge that cult members had been forced to enact a bizarre ritual, meant to enforce their "belief in the sanctity of Nature," in which plastic bags had been solemnly placed over their heads to symbolize their ultimate alienation from Nature. Of the bodies found in the burned-out ruins of the Swiss chalet, most wore plastic bags over their heads, pierced with bullet holes.

Marketers, familiar with the popular knee-jerk rejection of plastic, count upon it, often cynically, to sell goods. The New York Center for Plastic Surgery promises "Plastic Surgery That Doesn't Leave You Looking Plastic." MasterCard proclaims, "It's About Time Someone Took the *Plastic* out of Credit Cards." Why? Because, in a remarkable non sequitur, "In the real world, gold cards aren't status symbols. They're tools."

"Take Another Look at Plastic," a trade group calling itself The American Plastics Council exhorts us incessantly in TV and print ads. They take a different tack, accentuating all the positives about plastic. Like the fact that it's "strong, thin and lightweight, so it provides a lot of protection without a lot of additional heavy packaging." Like the fact that some plastics—though not many—are being recycled in bulk.

"Your New Carpeting May Already Be in Your Refrigerator," the ads soothingly say, featuring heartwarming tales of plastic bottles being turned into "toys, pillows, garbage cans, sailboat sails, even plastic 'lumber'"—right before our very eyes.

But the message, at least in some quarters, seems to be getting through. After decades in the doghouse, plastic has come back with a vengeance. Patagonia's EcoFleece and Wellman's EcoSpun, all-American products that turn soda-bottles back into a fibrous fabric that even die-hard naturalists are proud to wear into the deepest woods have, here in the '90s, made drecky synthetics respectable again.

We mold plastic. And plastic molds us. With remarkable resiliency and pliability appropriate to its protean nature, plastic has cunningly mutated back into our good graces in recent years. Philippe Starck's "zany cactus-in-a-pot" plastic toilet brush was dubbed by the design magazine *Metropolis* "a toilet brush with greatness." Fashion Week '95 in New York was "awash in synthetic glamour," writes Maureen Dowd of the *Times*. Of one model's plastic suit, the wearer raved, "When it gets dirty I can just wipe it down with Windex." Shades of Betsey Johnson. The clothes were made of metallic acetate, iridescent viscose, rayon fringe, plastic lace, polymicro nylon, "pleather" (what, pleather again?), Lurex. Everything but Pyrex.

Plastic now extrudes from the tube in ecologically fashionable fibers, in dissolvable films, in recyclable bottles, and even in purportedly green "bioplastics." These include polymers made from soybeans, cornstarch, and potatoes. Which, by the way, is nothing new in the field with the bright future. The first plastics were made from cellulose—pure vegetable fiber. Henry Ford made plastics from soybeans. In the twenties, the Germans made plastic from cow's blood from abattoirs.

Still, plastic gets no respect. Even when it deserves it. The people who most depend upon plastic are the poor, the displaced, and the dispossessed. Plastic is the refugee lifesaver. In Rwanda, villagers fleeing from ethnic violence fretted over having no goats, no soap, and no clay pots in which to carry water. But one displaced woman, standing beside her hastily erected hut made of sticks and covered with leaves and ferns, worried that the worst lay ahead. "When the rains come, we'll die," she cried. "We have no plastic sheeting."

In Africa, where half the population is under fifteen, "the plastic bucket has revolutionized the lives of Africans," writes Ryszard Kapuscinski.

You cannot survive in the tropics without water; the shortages are always acute. And so water has to be carried long distances, frequently dozens of kilometers. Before the invention of the plastic bucket, water was carried in heavy vats made of clay or stone. The wheel—and vehicles that use it—was not a familiar aspect of African culture; everything was carried on the head, including the heavy vats of water. In the division of household labor, it was the woman's task to fetch water. A child would have been unable to lift a vat. Acute poverty meant that few households could afford more than one vat.

The appearance of plastic buckets was a miracle. To start with, it is relatively cheap (although in some households it is the only possession of value), costing around two dollars. And it is light. And it comes in different sizes: Even a small child can carry a few liters. Now it is the child's job to fetch water. Flocks of children playing and bantering on their way to a distant well are common. What a relief for the overworked African woman!

In fact, the positive features of the plastic bucket are endless. Consider the queue. When water was brought in by cisterns, it was often necessary to queue up for a whole day. The tropical sun is relentless. You couldn't risk leaving your vat in your place in the queue and retiring to the shade. It was too expensive. Today, the bucket has taken the place of the individual in the queue. Today, the queue frequently consists of a long, colorful line of plastic containers, their owners waiting in the shade, off to the market, visiting friends.

Plastic. There's a great future in it. Think about it. Will you think about it?

Yes, *sir*.

2

Protoplastic

Interior view of the Great International Exhibition, London, 1862
(Courtesy of The New York Public Library Picture Collection)

The manufacture of cellulose nitrate-camphor containing plastics is essentially an imitative industry, and a forgery of many of the necessities and luxuries of civilized life. But unlike many forgeries, these plastics possess properties superior to those of the originals which they are intended to simulate. The existence of the industry ... depends upon how closely it can reproduce that which is beautiful and scarce, and hence much sought after and costly.

EDWARD CHAUNCEY WORDEN
NITROCELLULOSE INDUSTRY (1911)

The night Queen Victoria's husband Prince Albert died in December 1861, the bearded sexton at Windsor Castle—ordinarily an impassive man—burst into torrential tears. By the next morning, every shop in London had put up black shutters, while sizable crowds gathered outside the photographers' shops, gazing mournfully at the few portraits of the prince left unsold. By the end of the day, the nation's stock of black crepe was exhausted. Queen Victoria, donning her widow's weeds, would wear nothing but black for the next forty years.

For much of the year leading up to his untimely death, the genial German-born prince, a passionate patron of science and industry, had been deeply involved in planning London's second Great International Exhibition of 1862. Like its more illustrious predecessor, the Great International Exhibition of 1851, London's second Great International was to serve as a spectacular showcase for the progress recently achieved in the industrial arts and sciences not only by Great Britain but by every nation on earth. By inventing the modern world's fair—a great trade show where every nation could compete for the acclaim of the crowd—he endeared himself to his adopted country. Prince Albert made his mark as a scientific impresario, a man of the people who produced the gathered fruits of science and industry as events to be proud of, spectacular and democratic.

At twelve o'clock sharp on a sunny May Day, an honor guard of Grenadiers sounded a flourish of trumpets outside the sprawling Exhibition Building in South Kensington, appropriately located at the intersection of Prince Albert and Exhibition Roads. Contingents of Life and Horse Guards lined the broad avenue on both sides, gorgeous in state dress, their gilded helmets glinting in the spring sun.

Inside the nave, a swirling procession of princes, earls, marquises, and dukes, joined by "noble and royal ladies in gay and jewelled attire," marched "with measured tread" to a grand overture played by an orchestra composed of two thousand voices and four hundred instruments, followed by a chorale sung to an occasional poem composed by the poet laureate, Alfred Lord Tennyson.

Harvest tool and husbandry,
Loom and Wheel and Enginery
Secrets of the sullen mine
Steel and Gold and corn and wine.
All of beauty, all of use
That one fair planet can produce
Brought from under every star
Works of peace with works of war . . .

Outside the building, hundreds of thousands of spectators thronged the surrounding streets, hoping to make way on foot or by horse-drawn bus ("Leaving Every Ten Minutes for the Great Exhibition!") to catch a glimpse of the grand pageant in progress. The sole pall cast over the day's proceedings, the *Times of London* sadly noted, "was that to those who mourned the Prince's passing, the vast structure which the Prince intended as the region of his activity had become, by a strange and unlooked-for turn of fate, a mighty mausoleum to his memory."

By two that afternoon, the mighty mausoleum had flung open its doors to a broader public, the first of some six million visitors to descend upon the people's palace in droves throughout the coming year. The prince's legacy was best expressed by the division, at his suggestion, of the products of human endeavor into four basic categories:

1. Raw materials
2. Machinery
3. Manufactures
4. Fine arts

Eleven years before, at the opening of the first Great International at the Crystal Palace in Hyde Park, Prince Albert had hailed "the coming victory of Man over Nature," proclaiming with pride that "man is approaching a more complete fulfillment of that great and sacred mission—to conquer Nature to his own use."

During the coming decade—the 1850s—cotton and coal production in England would double. Iron production and railway mileage would nearly triple. As the wheels of industry spun overtime, hoisted into high gear by the boundless pressure of burning coal, boiling water, and pumping pistons, nature's bounty, on which mankind had for so long relied, was being rapidly replaced by a more abundant surplus liberated from the earth's stock of raw materials by the steam-driven machine.

Still, of the six million visitors drawn to this industrial ur-Epcot, no more than a handful were attuned to the finer subtleties of a quieter, less celebrated revolution in progress, a great leap forward not of "Harvest tool and husbandry, Loom and Wheel and Enginery" but of chemistry.

To be sure, one official guidebook to the 1862 exhibition dutifully marveled: "From a block of coal, we may obtain not only light and heat, but also fragrant essences and brilliant dyes." These "fragrant essences and brilliant dyes," derived from coal, had imposed the few flashes of color to be seen in the drab, sooty industrial landscape of Britain. And in the space of a few years, they exerted so profound an impact on popular culture in Europe that the 1850s were destined to be defined by a dye.

The Mauve Decade

In 1802, homes and factories throughout Great Britain began to be illuminated by natural gas derived from coal deposits. Over the next twenty years, as gas companies incorporated in every major British city, piping this cheap, relatively reliable form of light from the filthy Yorkshire mines to everyone and anyone who could afford it, the wildly proliferating gasworks produced a noxious spew of unsavory by-products that turned the verdant pastures of once rural England a sooty, sulfuric brown.

The most common of these, and by far the filthiest, was a worthless gunk called "coal tar." It clogged up the flues of the gasworks, and appeared to be good for nothing better than preserving—in the form of pitch, or creosote—the millions of wooden sleepers being churned out to feed the railroad-building mania that swept Britain during the 1840s.

Coal tar would never have amounted to more than a tinker's damn if a German-born chemist, August Wilhelm von Hofmann, had not had the insight to conduct an in-depth investigation of some of its more intriguing properties. These featured a capacity to be distilled into a dazzling range of potentially valuable chemical compounds.

At the behest of Prince Albert, a firm believer in the industrial value of scientific research (a governing precept in his native Germany) Hofmann, a disciple of the legendary Justus von Liebig, founder of the first chemical research laboratory at Giessen, agreed to take up the first directorship of London's new Royal College of Chemistry. Hofmann was theoretically brilliant, but all thumbs in the lab. So he hired a bright eighteen-year-old assistant, William Henry Perkin—the son of a wealthy builder—to lend him a hand.

Toward the end of the 1840s, as malaria continued to scourge the tropics—posing one of the greatest impediments to British imperial interests in the equatorial region—Hofmann speculated to Perkin that the antimalarial drug quinine (then naturally extractable, at great cost, from the bark of the South American cinchona tree) might be derivable, perhaps even synthesizable, from common coal tar.

While on Easter break, Perkin set to work on this sticky problem at his parents' luxurious home at Harrow. Using as his starting point the coal-tar derivative aniline, he oxidized a pinch of it in combination with potassium dichromate before dissolving it in alcohol.

Rather than produce a synthetic quinine, he ended up with a khaki-colored stew with no redeeming pharmaceutical qualities. But, by a strange twist of fate oddly common in those early days of trial-by-error chemistry, Perkin happened to spill some of the muck out of a beaker onto his lab bench. After scrubbing the mess up with a clean cotton rag, he was astounded to see the rag come away stained an elegant lavender. On a whim, he dipped some strips of silk into the spill, turning them a deep purple rich enough to dye the hem of Tiberius's toga.

As Perkin well knew, the Romans had reserved the color purple for its emperors precisely because it was so hard to extract from a viscous secretion exuded by the rare shellfish Purpura. In the centuries since, Continental colorists had obtained purple from plants, but the process was costly and cumbersome. Perkin had ironically—not to mention accidentally—derived one of the brighter colors of the rainbow from the same soot that rained down over grimy London.

Impressed by the intensity of his synthetic hue, Perkin laundered the silk strips in soap and water and hung them out to dry in the sun. The deep purple held fast. He had the new dye checked out by the venerable Scottish firm Pullars, which took pleasure in informing him that it "showed an enviable color fastness in sunlight, and mordanted well to silks"—that is, held colorfast.

As the son of a prosperous businessman, Perkin had a hunch that if aniline mauve could undercut its natural French rival *pourpre français*—derived from an extract of mallow blossom—he stood to rake in a fortune. With his father's encouragement (and much to his mentor Hofmann's regret) the promising Perkin abandoned a stellar academic career to go into the dye business with his brother.

The pale purple silk costumes that soon issued forth from British silk mills became so wildly popular among fashionable ladies that the 1850s became known as "the Mauve Decade." But mauve was just the first of an artificial rainbow of synthetic colors to come spilling out of the brimming vats of W. H. Perkin & Sons' bustling factory at Harrow. Perkin had single-handedly founded not only the synthetic dyestuffs industry but also laid the groundwork for the modern pharmaceutical trade. Perkin—now a millionaire—and the new breed of "industrial chemists" his infant industry rapidly spawned soon found that certain synthetic compounds affected organic substances, striking directly at germs and other disease agents while leaving the live tissue unscathed. While Perkin and his disciples in the British dyestuff industry concentrated on tinting textiles, industrial chemists in Germany continued to try and solve the problem Hofmann had originally posed to Perkin: the synthesis of chemical equivalents of rare and powerful herbal drugs from coal and other organic compounds.

By the time of Prince Albert's death, mauve enjoyed its first flush of success as an alternative to black for mourning. Before long, its price had fallen so low that the British Post Office dyed penny stamps with it.

Protoplastic

An official guide to the 1862 Great International Exhibition noted two remarkable curiosities. The first was quite literally striking: "a new match which could not be ignited by friction alone," soon to gain global fame as the lifesaving "safety" match. The second, rightly regarded by the author as "among the most extraordinary substances shown at the Fair," was said by its inventor to be "the product of a mixture of chloroform and castor oil, which produces a substance hard as horn, but flexible as leather, capable of being cast or stamped, painted, dyed or carved and which can be produced in any quantity at a lower price than Gutta Percha." This was Parkesine.

Anyone with even a passing familiarity with the sterling reputation of Alexander Parkes—best known to the public as the ingenious man who had figured out a way to "cure" rubber at cold temperatures—would have been hard put not to imagine that Parkesine was another one of the new modified rubbers then generating so much excitement with the general public.

The son of a lockmaker, this extraordinarily prolific metallurgical and chemical genius had served a long apprenticeship as a casting specialist at two Birmingham brass foundries before earning some eighty British patents while taking the time out to father twenty children by two wives—the second a close friend of his eldest daughter's.

An explanatory leaflet displayed alongside the Parkesine booth promoted the astonishing versatility of his new "vegetable product":

> In the case are shown a few illustrations of the numerous purposes for which Parkesine may be applied, such as MEDALLIONS, SALVERS, HOLLOW WARE, TUBES, BUT-

TONS, COMBS, KNIFE HANDLES, PIERCE and FRET WORK, INLAID WORK, BOOKBINDING, CARD CASES, BOXES, PENS & PENHOLDERS.

It can be made HARD AS IVORY, TRANSPARENT or OPAQUE, of any degree of FLEXIBILITY, and is also WATER-PROOF; it may be made of the most BRILLIANT COLORS, can be used in the SOLID, PLASTIC or FLUID STATE, may be worked in DIES and PRESSURE, as METALS, it may be CAST or used as a COATING to a great variety of substances; it can be spread or worked in a similar manner to INDIA RUBBER, and has stood exposure to the atmosphere for years without change or decomposition. And by the SYSTEM OF ORNAMENTA-TION patented by HENRY PARKES (the inventor's brother) in 1861, the most perfect imitation of TORTOISE SHELL, WOODS, and an endless variety of effects can be produced. . . .

SPECIMENS of which may be seen in the CASE . . .

> › PATENTEE AND EXHIBITOR
> ALEX. PARKES, BIRMINGHAM

Parkesine was the first man-made plastic: an organic material that can be molded under heat and pressure into a shape that it retains once the heat and pressure are removed. Rubber, amber, and glass are natural plastics—moldable under heat and pressure into a variety of shapes. Parkesine represented the first attempt to dupli-cate these materials by chemical manipulation—in the lab.

Surrounding Parkes's Parkesine booth in the Eastern Annex, in the Vegetable Substances section of the exhibition, lay a profusion of fourteen thousand exhibits, an overwhelming number of which were fashioned from the new vulcanized rubber.

From the prominent firm of Thomas Hancock & Sons: a collec-tion of "Vulcanized India rubber braces, and other surgical devices and appliances."

From the Manchester firm of Clark & Co: "India rubber for the waterproof lining of packing cases, attested to by Lloyds for certifi-able moisture protection."

From Charles Macintosh, of Glasgow: "Waterproof clothing, including military, naval, travel and sport, surgical and hospital instruments . . . "

From George Spill of Hackney Wick: a "waterproof vegetable fabric, mineralized, of Superfine India Rubber Cloth, weighing just six ounces, warranted to withstand heat under 400 degrees F."

Obscured, if not lost, among this splendid array of rubber objects was Vegetable Manufactures Exhibit #1112, "Various Articles of Patented Parkesine of different colors: hard, elastic, transparent, opaque, and waterproof."

But what was Parkesine? Parkes preferred to say what it was not: rubber. And Parkes touted what it could do—anything that rubber could, only better. And cheaper.

Nature's Plastic

In 1496, on his second voyage to the New World, while visiting the Caribbean island of Hispaniola (now the Dominican Republic and Haiti), The Great Navigator had looked on with delight as two teams of native village men played a hard-fought soccerlike game with a ball "made of the gum of a tree, which though heavy, could fly and bounce better than those filled with wind in Spain."

In 1611, Hernán Cortés's lieutenant in Mexico, Juan Ribera, observed the Aztec nobility playing an even more violent version of the same game Columbus had seen on Hispaniola. The players used only elbows and knees (not hands or feet) to hit a large ball that Ribera described as being molded from "the juice of a certain herb, hardened by heat, which being struck upon the ground softly, would rebound incredibly in the air."

The Europeans had never before seen a solid object that could bounce like an air-filled leather ball. But rubber had other uses than for games. Captive Mexican Indians soon taught their Spanish overlords how to waterproof cloaks by spreading the milky rubber sap on cotton cloth, and to fashion inflatable bags and bladders, and "syringes that could be squeezed and snapped back to their original shape." As well as to pour rubber sap on the feet, let it dry, and peel it off in long strips, which could be glued together to make custom-cut waterproof boots.

In 1735, Charles Marie de La Condamine, a French nobleman and leader of an expedition sponsored by the French Academy of Science charged with taking precise measurements of the earth at the equator, brought back to Paris from Peru some samples of the marvelous substance Peruvians called *caoutchouc*. "From this 'weeping wood' may be made bottles which are not fragile, and waterproof boots," La Condamine reported excitedly to his fellow academicians. "And hollow balls which flatten out when pressed and regain their former shape when released!"

Intrigued by the one-piece waterproof boots the Peruvians made by smoking the raw sap until it attained the appearance and consistency of leather, La Condamine's colleague Fresneau proudly fashioned a pair of waterproof boots that quickly became the talk of fashionable Paris.

In 1770, the eminent British chemist Joseph Priestley (best known as the discoverer of oxygen) observed that *caoutchouc* was "excellently adapted to the purpose of wiping from paper the unwanted marks of a black lead pencil." It certainly beat bread crumbs, the best known candidate for the job, at rubbing out errors. Priestley called *caoutchouc* "India rubber" because (1) it came from the West Indies and (2) it made a good "rubber."

Unfortunately, those brave entrepreneurial souls who brought rubber home to Europe from the balmy tropics with the idea of making a killing on it in the Old World were in for a rude shock. Rubber frequently failed to act rubbery at all in the more temperate climate of Europe. On hot summer days, it turned intolerably sticky and tacky to the touch. In the depths of a cold European winter, it became brittle, hard, and lifeless. These natural defects would keep rubber on the industrial sidelines as a curiosity and a novelty for the next several centuries, until some remarkably clever fellow figured out some way of "curing" it.

The Mack and the Masticator

Fifty years later, a thrifty Scottish dye maker, Charles Macintosh—who used the coal-tar derivative ammonia in his dye-making factory—made something useful out of this tropical by-product. For a few pence, Macintosh contracted to buy all the coal tar produced by the Glasgow Gas Works, which was more than happy to get rid of the gunk.

After separating out the ammonia in the process of converting tar into pitch, poor Macintosh found himself with too much of another worthless distillate on his hands to suit his frugal tastes—a sharp-smelling waste product called naphtha. Determined to make this naphtha good for something, Macintosh tried—on a whim—immersing some raw rubber in it. To his dour delight, he obtained a liquid rubber solution that could be brushed on cloth, like a varnish. In 1821, he began coating sheets of cotton canvas with rubber solution and pressing two sheets together to form a rubber sandwich. By the fall of 1823, British gentlemen proudly sported waterproof "Macks" in the densest London fogs.

Their fashionable American cousins were less well served by the spirit of progress. Hoping to get the jump on Macintosh's American patents, Yankee manufacturers rushed to market with cheap rubberized cotton knockoffs. The resulting raincoats melted into gooey capes during hot American summers, particularly in the sultry South.

Coincidentally, in the same year (1820) that Macintosh perfected his naphtha-ized rubber, Thomas Hancock, a London-based coach maker and operator, trained his own eye on the same raw material with an equivalent goal in mind: producing his own line of waterproof clothing. As a coach-line operator, Hancock had good reason to keep his long-suffering passengers from getting soaked to the skin while driving down muddy English country roads.

At first, Hancock experimented with immersing strips of raw rubber in turpentine, which yielded solutions that dried into useless scum. But being more mechanically than chemically adept, he soon ceased puttering around in the lab and instead tooled up in his machine shop a gadget he called—for security purposes—a "pickle."

Hancock's "pickle"—a plain wooden barrel studded on the inside with iron teeth, which meshed with more iron teeth mounted on an iron roller—helped solve one of rubber's most egregious defects: Though freshly cut surfaces of rubber adhered well together, once the cut surfaces were exposed to air for any length of time, they no longer joined easily.

Expecting to shred the rubber into disparate parts, Hancock was pleasantly surprised to find that after opening his "pickle," vigorous cranking had indeed shredded the rubber. But if allowed to set awhile, the shredded parts in time congealed into a solid, impermeable mass, from which Hancock was able to slice sheets that miraculously held together. After patenting his "pickle,"

Hancock was prepared to reveal its true name and nature: the Masticator.

Goodyear's Blimp

While Hancock and Macintosh were fabricating rubberized cloth in England, a financial panic in the United States wiped out a flourishing hardware business in New Haven, Connecticut. After being forced by debtors to close his father's shop, the young Charles Goodyear hit the road, determined to make his name as an inventor.

For just as long as he could remember, Goodyear had marveled at "the wonderful and mysterious properties" of rubber. As a strange, lonely young boy growing up on his father's farm outside New Haven, he had been given a hot water bottle to keep his bed warm through the cold New England nights. He adored that soft rubber bottle, so soft and squeezable, so consistently warm to the touch, so reliably and pleasantly pliable.

In New York, he promptly paid a call on the managing partner of the Roxbury India Rubber Company, America's oldest and largest rubber enterprise, with an idea for improving the brass airtight valve on one of the company's most profitable products: an inflatable life vest.

Charles Goodyear
(Courtesy of The New York Public Library Picture Collection)

The managing partner of the company turned down Goodyear's proposal for improving the airtight valve on the dismal grounds that there was little point in trying to sell rubber products until some bright fellow—who knows, it could even be Goodyear—came along who knew how to make not a better valve but a better rubber.

Fixating on this chance remark uttered in a moment of commercial despair, young Goodyear found his life's mission. Undaunted by his total ignorance of classical chemistry—not to mention seed money—Goodyear embarked on a quixotic quest to save rubber from its own natural faults. Frequently running out of money and sinking deeper and deeper into debt, he spent so much time in and out of debtors' prisons that he caustically referred to them as his "hotels." He was occasionally able to persuade a sympathetic warden—whom he would regale with tales of his rubber wizardry—to let him dabble with his pots of chemicals as his work detail in the prison kitchen.

Goodyear tried mixing crude rubber with just about everything he could think of: witch hazel, castor oil, ink, and—in a pinch—chicken soup and cream cheese. In the kitchen of a small cottage on the grounds of one of his "hotels," Goodyear artlessly stumbled upon his first significant insight into raw rubber's mysteries when he obtained some moderately successful results by treating rubber with nitric acid vapors laced with sulfuric acid. Nitric acid appeared to dramatically improve rubber's heat resistance, which was half the battle. Now, all he had to do was figure out some way to cure rubber's stiffness when cold, and he was home free.

After being bailed out of debtors' prison in 1837 by a small rubber manufacturer, Goodyear made rapid progress until, in the wake of another stock-market crash, his jittery backer pulled the plug. The starving inventor was forced to camp out on the grounds of his defunct factory on Staten Island, a wild outpost in New York Harbor, frantically fishing to feed himself until a friend at the post office came to his rescue by offering him a contract for nitric-acid-cured rubber mail bags.

It looked as if Goodyear's fortunes had finally taken a turn for the better. But alas, as luck would have it, on a blisteringly hot summer's day in 1838, while inspecting his first batch of precious mail bags that had spent a few days in an unventilated storage room,

Goodyear was crestfallen to find that in the heat they had crumbled into piles of pellets. Subsequent analysis revealed that the nitric acid had cured only the surface of the rubber, leaving the interior as vulnerable to heat as ever. After hearing the bad news, the post office canceled Goodyear's contract, and Goodyear went back to cutting bait.

In 1839, still no closer to his goal, Goodyear gamely commissioned a rubber suit, strictly as a publicity gimmick for the wonders of rubber. He had become a rubber maniac. As one of his acquaintances said at the time, "If you meet a man in the street who has on an India rubber cap, stock, coat, vest and shoes, with an India rubber purse in his pocket with no money in it—that's Charles Goodyear."

Remarkably enough, Goodyear was not in prison when, in his fortieth year, on the grounds of a small Massachusetts rubber company that had granted him space to conduct his usual round of slapdash experiments, he accidentally spilled some raw latex mixed with sulfur on a hot stove. He heated it up to a fever pitch, but rather than melting into goo as ordinary raw rubber would have done, Goodyear's sulfurized rubber charred evenly, like a piece of leather.

With trembling fingers, he nailed the bit of charred rubber to a board and hung it outside in the bitter New England wind. Instead of stiffening up, it stayed pleasantly pliable—just like his old hot water bottle.

Though woefully ignorant of the chemical processes involved, Goodyear had spilled himself into the best possible chemical method of turning raw rubber from a tangled mass of chainlike molecules into an orderly arrangement of molecules "linked" in latticelike strands. The term later used would be polymers.

After filing for an American patent on his sulfur-treatment process, Goodyear had no money left to file for a British patent. Instead, he resolved to approach the flourishing firm of Charles Macintosh & Co.—which had grown wealthy on its patented "Macks"—to see if they might back him in a British-American joint venture. Confident of securing the American market for his "cured" rubber, Goodyear was not averse to licensing the British rights—in fact, he was typically desperate for capital.

Having worked out this scheme down to the slightest detail, he appointed a young Englishman, Stephen Moulton—who was just preparing to return home after spending several years in the United States—as his official British representative. In April 1842, Moulton set sail for England with a few samples of Goodyear's "improved" rubber concealed in his bags. After docking and taking the first train to London, Moulton lost no time in showing his samples to Charles Macintosh & Co., where the redoubtable Thomas Hancock had recently landed as managing director.

To put it mildly, Hancock found Moulton's samples intriguing. While advising Moulton (or so he later claimed) that he should tell Goodyear to take out a British patent (so that he and his partners could judge the value of Goodyear's process impartially, without compromising the American's closely guarded trade secret) he asked Moulton to leave some of his samples with him as a sign of good faith.

To Goodyear's endless regret, Moulton did just what he was told. Given the opportunity to peruse Goodyear's rubber at his leisure, Hancock detected a pale sprinkling of powder on the surface of the rubber. It smelled unmistakably of sulfur. At first convinced that Goodyear had deliberately rubbed sulfur onto the surface to throw him off the scent, Hancock spent the next few weeks ensconced in his lab, working around the clock trying to tackle the problem himself.

After proving to his own satisfaction that sulfur was the key to this new and improved rubber's remarkable resiliency, Hancock quickly and quietly applied for a British patent—just two months before Goodyear belatedly filed. At the suggestion of an artist friend, a "Mr. Bockedon" known for his fondness for classical allegorical scenes, Hancock called his new process "vulcanization," in honor of the Roman armorer and blacksmith to the gods.

While Hancock forged full steam ahead to capitalize on his British patent, an enraged, increasingly desperate Goodyear (who still retained American patent rights) retreated into a rubberized dream world, lost in contemplation of ever more fantastic uses for his miraculous "elastic metal" and "vegetable leather." While suing Hancock for patent infringement, he placated himself by filling ream after ream of lab notebooks with vivid, imaginative drawings of rubber life rafts, rubber bands, rubber wheelbarrow tires, and— bizarrely—rubber baptismal pants.

In 1846, determined to counter Goodyear's transatlantic accusations of fraud and patent infringement, Thomas Hancock demonstrated the versatility of the new vulcanized rubber—and his own considerable flare for publicity—by outfitting—to much fanfare—Queen Victoria's carriage with solid rubber tires. Oddly enough, though Goodyear was infatuated with inflatables, the "pneumatic" rubber tire would not be invented until the 1860s—and would then have to wait fifty years to be "reinvented" by J. B. Dunlop, a Belfast veterinarian in search of a softer ride for his son's tricycle. Smoothing the royal ride over London's rough-and-tumble cobblestoned streets became the rubber magnate Hancock's gift to a grateful nation.

Goodyear retained the great Daniel Webster, at an astounding wage of $15,000 for two days' work, to argue the British patent suit on his behalf. With the patent suit still unresolved, Goodyear planned to launch his greatest assault on Hancock's claims by mounting his greatest publicity stunt to date, to take place on Hancock's home turf. He borrowed extravagantly beyond the limits of his available funds to underwrite a magnificent Vulcanite Court at the Great Exhibition of 1851, where every object on display, including the walls and roof of the vaulted room, was fashioned from vulcanite rubber. Hard rubber canes, musical instruments, and his personal favorite, inflatable furniture, were overseen by huge rubber balloons, some six feet in diameter, suspended from the ceiling on guy wires. Presiding over the show like a great whale was a huge rubber life raft, equipped with Goodyear's first invention—an improved airtight brass valve.

The Gutta-Percha Telegraph

The same year (1846) that Hancock put tires on Victoria's carriage, Dr. Peter Montgomerie, a British surgeon practicing in Madagascar, observed with quickening commercial interest a group of natives molding smooth, shiny knife handles out of a dark brown material they called "gutta-percha." When he asked where the stuff came from, he was told that it had been extracted from the sap of the Paraquium tree, indigenous to Malaysia. They softened it in hot water, pressed it by hand, and molded it into a rock-hard mass after cooling.

By Jove! Montgomerie exclaimed. Wouldn't this hardy stuff make a slap-up insulator for telegraph wires? The recent rapid expansion of international telegraphy had been hampered by one obstacle: the lack of a suitable insulator to coat submarine lines with some "dielectric"—electric nonconducting—material tough enough to withstand the corrosive effects of salt water. Even in vulcanized form, rubber was not sufficiently corrosion resistant to justify the massive capital investment involved in laying miles of copper cable along the seabed. But gutta-percha turned out to be not merely resistant to sulfur (which rendered it unvulcanizable) but also to salt, most acids, and electrical current.

Gutta-percha was just about perfect. But how to coat the cable? The British inventor H. Bewley in 1845—a year before Montgomerie's "discovery"—had patented a machine called an "extruder," which pumped molten rubber through an orifice into a pipe, forming hollow tubes of any desired length. In 1850—a year before the opening of the first Great International—the world's first submarine telegraph was laid beneath the English Channel, running twenty-five miles between Dover and Calais, reliably sheathed in gutta-percha.

Guncotton

In 1846, Christian Friedrich Schönbein, a professor of chemistry at the University of Basel, Switzerland, was distilling nitric and sulfuric acids in his kitchen when he knocked over a flask containing his experiment. As the corrosive liquid spread all over his wife's clean linoleum floor, he grabbed the first thing he could lay his hands on—a cotton apron—to clean up the mess. After washing out the apron with hot water, he hung it by the hot stove. Instead of drying, the acid-impregnated cloth burst into flame, before vanishing in a pale puff of smoke.

Schönbein, best known for discovering ozone, had accomplished that feat by endlessly experimenting with oxidization—a chemical reaction that adds oxygen to or subtracts hydrogen from chemical compounds. In the course of those investigations, he learned that a mixture of nitric and sulfuric acid was a powerful oxidizing agent. As a way of producing ozone, Schönbein nitrated (treated with nitric acid) just about every organic compound he could lay his hands on. The most promising of these was cotton, one of the purest natural forms of cellulose on earth—the fibrous material of which the cell walls of all plants and living things are made.

Schönbein, however, was not the first chemist to nitrate cellulose; that honor goes to two French chemists, Théophile Pelouze and Henri Braconnot, who a few years earlier had created the first nitrocellulose compounds. Pelouze called his version "pyroxyline"; Braconnot coined the term "xyloidine."

Schönbein's legacy to nitrocellulose chemistry was adding sulfuric to the nitric acid as a catalyst to enhance the reaction. While Pelouze's and Braconnot's partly nitrated cellulose compounds burned brightly if brought into contact with fire, Schönbein's more fully nitrated cellulose exploded with a fury so awesome that it made black gunpowder seem as insipid as a child's firecracker. Within a few weeks of his timely spill, he wrote excitedly to his good friend Michael Faraday, the discoverer of electromagnetism:

Feb 27, 1846

I have of late performed a few chemical exploits which enabled me to change *very suddenly, very easily and very cheaply* common paper in such a way as to render that substance exceedingly strong and entirely waterproof. I send you a specimen of a transparent substance which I have prepared out of common paper. This matter is capable of being shaped out into all sorts of things and forms and I have made from it a number of beautiful vessels.

At the same time, he sent samples of his unusual transparent paper to an old colleague, J. C. Poggendorf, editor of the Swiss science journal *Annalen der Physik und Chemie*, who kindly replied:

Your glass-like paper is splendid. I hope you can make it thick enough to use for window panes. . . . Might it not also make a good replacement for ordinary paper for bank notes?"

Transparent banknotes (particularly Swiss ones) and beautiful vessels were all very well, but Schönbein was far more interested in determining the commercial value of the awesome violence he had serendipitously stumbled upon. Though known around Basel as "energetic, quick-moving, short, stout, and eminently friendly," the affable Schönbein had discovered within himself latent bellicose tendencies. The more he worked with guncotton, the more he became convinced that he had stumbled upon not just a gold mine but a land mine.

To Faraday he coyly confided:

I am enabled by this process to prepare, in any quantity, a matter which, next to gun-powder, must be regarded as the most combustible substance known. . . . I think it might be advantageously used as a powerful means of defense or attack. Shall I offer it to your government? PS: I must of course beg you to keep it entirely to yourself and consider this a confidential communication.

Faraday appears to have kept his mouth shut. But Schönbein couldn't resist rushing his potentially earth-shattering results into print. "This explosive cotton wool," he openly boasted, "will soon entirely replace gunpowder."

Schönbein's optimism was hardly misplaced. Common "black" gunpowder—imported from China into Europe in the 1300s—suffered from one major defect: It produced so much thick, billowing smoke that every significant artillery exchange evoked Matthew Arnold's ironic image of "ignorant armies clashing by night." Guncotton, which produced substantially less smoke and packed a far greater bang for its buck, was soon being referred to in hushed tones around military installations across Europe as the elusive "smokeless" gunpowder.

Rather than disclose his production methods in scientific journals, Schönbein took the precaution of applying for patent protection. He sold the formula for guncotton outright to the Austrian government, and in February 1846—just as he was dashing off his confidential missive to Michael Faraday—traveled to England to sign a contract with English patent agent John Taylor to license the production of guncotton in Britain.

In France, Germany, and Russia, Schönbein had no luck selling his formula. Instead, the French minister for war appointed a special investigatory commission to evaluate nitrocellulose as an alternate gunpowder and to figure out some way of duplicating Schönbein's results. The duke of Montpellier, the commission's head, hired Pelouze, the first man to nitrate cellulose, to work around the clock in a locked laboratory, in a mad arms race to recreate Schönbein's reaction before anyone else could. Before the year was out, England, Russia, Austria, and Germany had all set up comparable commissions, composed of armaments experts and chemists, seeking to develop at breakneck speed the nineteenth-century equivalent of the atom bomb.

In July 1846—the year of Montgomerie's gutta-percha, Queen Victoria's rubber tires, Parkes's cold vulcanization, and Schönbein's spilled flask—the British Ministry of War invited the increasingly combative Schönbein to test his new explosive at the ballistics proving ground at Woolwich. There, he became the first man to fire a cannon charged with guncotton, ably demonstrating that "guncotton could be used as a propellant for artillery and that practically no smoke was formed when the gun was fired." In attendance at the demonstration, at the invitation of the British army, was the redoubtable Alexander Parkes.

In England, the patent agent John Taylor had entered into an exclusive agreement with the firm of John Hall & Sons to produce guncotton for commercial purposes as a blasting agent. Unfortunately, guncotton's greatest strength—its extraordinary volatility—was also its worst weakness. On July 14, 1847, the Halls' plant at Faversham blew up, killing twenty workers. Before the year was out, similar explosions leveled labs at Vincennes and Bouchet in France, and wreaked havoc at armaments works in Russia and Germany.

Upon the stern recommendation of a hastily convened imperial commission, the czar promptly banned the manufacture or importation into Russia of this obviously uncontrollable high explosive. By 1850, a French commission found that despite guncotton's admittedly enormous potential, "under the present circumstances there is no use in continuing these experiments."

Collodion

In that landmark year of 1846—gutta-percha, Queen Victoria's carriage, Goodyear's patent fight, Schönbein's guncotton—Théophile Pelouze, professor of chemistry at the Collège de France and the first man to make cellulose nitrate, accepted into his private laboratory as an apprentice chemist the twenty-four-year-old self-styled artist and "revolutionary poet" Louis Ménard.

Pelouze was still struggling around the clock in his locked laboratory to duplicate Schönbein's reaction so that the French military authorities could produce guncotton as a ballistic material without having to pay Schönbein. Pelouze promptly put the young Ménard to work on one tiny aspect of that problem: the treatment of nitro-cellulose with a wide variety of solvents.

Ménard's true vocation was not as a chemist but as a revolutionary poet and painter who preferred lounging around the Latin Quarter with subversive elements—many of whom would become prime movers in the impending revolution of 1848—to fiddling in the lab. While Paris's café society was abuzz over the second attempt to assassinate the besieged King Louis Philippe, and while the inflammatory socialist economist Pierre-Joseph Proudhon churned out revolutionary pamphlets by the ream, Ménard, author of the political polemic *Le Prologue de Revolution* (regarded by the monarchical authorities as unduly incendiary) quietly achieved a subtler revolution in Pelouze's lab.

On November 9, 1846, at a specially convened meeting of the French Academy of Sciences, Pelouze announced that his student Ménard had created a new substance he called "collodion," from the Greek root for "adhere." Ménard had discovered that a particular combination of solvents—ether and alcohol—could liquefy nitrated cellulose into a smooth, clear, gelatinous lotion, in which it could be said to be in a "colloided" state. After being exposed to air, the "colloided" cellulose nitrate quickly dried into a tough, clear, transparent film, the possibilities of which appeared practically limitless—it could be a varnish, a lacquer, a waterproof coating for cloth, perhaps even a moldable solid.

Ménard's artistic inclinations steered him in precisely the wrong direction for exploiting his only enduring creation. Convinced that this marvelous new clear hard material would ultimately achieve its greatest utility as an improved painter's varnish—a modern counterpart to the legendary lost varnishes of the Italian Renaissance—poor Ménard discovered, to his dismay, that collodion's powerful solvent action blurs oil colors and whitens in contact with air, dashing its chances of ever becoming a lacquer capable of preserving paint for posterity. Deeply disappointed at failing the cause of art, he shed his remaining interest in chemistry and returned in frustration to manning the barricades of Paris. He was arrested in France, made a quick break for England, and after returning home three years later when the coast was finally clear, never set foot in a lab again.

It was left up to a New England Yankee and Harvard medical student, J. Parker Maynard, to find the first truly useful and peaceful application for Ménard's collodion. In 1848, Maynard published a ground-breaking article in the highly regarded *Boston Medical and Surgical Journal* entitled "The Original Application of a Solution of Cotton to Surgery," which he quickly followed up with a slim volume (privately published) extolling the virtues of collodion as just what the doctor ordered for healing the gaping wounds caused by the explosive shells guncotton had been designed to propel. When spread on an open wound or even a mild abrasion, collodion formed an airtight, watertight, nontoxic "seal" that allowed the skin to heal cleanly underneath. It quickly became the dressing of choice for cut, chapped, chafed, or abraded skin, and nipples cracked by breast-feeding. Patent-medicine merchants began churning the film out under colorful names: Peerless Peeling Plaster, Dr. Clarke's Cut Cure, Heal All, Seal Skin, Celacut, Wizard Corn Cure, Kaposi's Wart Collodion.

Photo-Synthesis

In 1851, just prior to the opening of the first Great Exhibition, an English sculptor, Frederick Scott Archer, published an account of the successful results he had obtained by using Maynard's collodion as a support for the light-sensitive emulsions used in photography. Archer's "Wet Plate Collodion Process" vastly improved upon the calotype invented a quarter century before by William Henry Fox Talbot, of Lacock Abbey in Wiltshire, and the daguerreotype simultaneously developed by Louis Daguerre, a French scenic painter and diorama artist. Daguerreotypes and calotypes were unique photographic objects—one-shot works of art. But because Archer's collodion Archergram could be used to print any number of identical positive images on photosensitized paper, it marked the advent of photography as a duplicative, mass-market art form comparable to printing.

Being an artist and a gentleman, Archer freely published his findings in a British photographic journal for all the world to benefit from, if it so desired. But Archer's largesse was met with acid indignation by William Henry Fox Talbot, who claimed worldwide rights to virtually all silver-based photographic processes on the basis of his original patents. After gracelessly enduring the three years of fierce public outcry that greeted Talbot's spiteful blocking of the Archergram, Talbot withdrew his suit. For the next thirty years the Archergram utterly dominated photography. But its inventor was destined to die penniless a few years after winning his moral victory over Talbot.

Plastic Mass

In the years leading up to the Great International of 1862, Alexander Parkes became intrigued by collodion. He spent a fair amount of experimental time trying to devise some method of employing it not only as a varnishlike finish to be spread on a glass plate but as a substitute for the glass plate itself. In 1855, after being issued a British patent for "collodion lacquer"—a potential replacement for the costly natural product derived from the secretions of the Asian "lac" beetle—Parkes conceived of casting collodion in thick slabs. If he could only make collodion thick enough, photography could be liberated from its slavery to fragile, cumbersome glass and opened up to the mass market.

Parkes never did succeed in casting collodion in thick enough slabs to compete with glass. If he had, there might never have been a Kodak, and George Eastman may have remained an obscure bank clerk. But Parkes, who though brilliant was easily diverted, became convinced that collodion's long-term potential far exceeded the then limited world of photography. As the wheels of industry turned faster and faster and the demand for mass-produced products grew ever larger, Parkes increasingly saw collodion as a medium with vast, untapped potential—the makings of a true "plastic mass."

By "plastic mass" Parkes meant a material that could be molded and cast into an infinite number of identical objects by machine, as opposed to a material—like wood or stone—that needs to be carved or shaped by hand one piece at a time. This matter of molding went straight to the heart of the Industrial Revolution, because a moldable material is ideally suited to mass manufacturing. The most exciting thing about vulcanized rubber was that it could be machine molded into such an astoundingly wide range of products, from elaborately "carved" walking sticks to ornate picture frames, without having to resort to skilled artisans hand-tooling each article as it rolled off the line. Vulcanized rubber's greatest drawback was its high cost, because the raw material itself still had to be laboriously gathered by hand. Not to mention that a cartel of Amazonian rubber barons controlled the world's rubber supply and were not inclined to let its price drop below a level guaranteed to yield them a staggering profit.

In 1847, Parkes outfitted a laboratory at Elkington, Mason & Co.—the Birmingham brass foundry where he had been employed for twenty years—exclusively for collodion experiments. Over the fifteen years leading up the 1862 exhibition, Parkes by his own account devoted "many years of labor and thousands of experimental trials to applying to peaceful industrial purposes a material hitherto only used for military, blasting and photographic purposes."

He was determined, above all, to perfect and patent a process by which he could turn out his "plastic mass" in tons—not pounds—per month and at a cost, as he advertised at the Great Exhibition, "cheaper than hard rubber or gutta-percha." Unfortunately, the solvents commonly used to prepare medical or photographic collodion were far too expensive for mass production. In place of such solvents, he tried mixing his nitrated cellulose with a variety of vegetable oils, to maintain his compound's pliability—and viability.

Soaring on his success at the 1862 exhibition—where Parkesine was awarded the coveted bronze medal out of the fourteen thousand exhibitions in its section of "Vegetable Manufactures" for "ingenuity of invention"—Parkes delivered a stirring lecture to London's Society of Arts. It was all part of a well-orchestrated fund-raising campaign to drum up sponsors for Parkesine's commercial production. An inspired advocate for his own cause, before his fellow members of the society he made the arresting claim that "given the proper machinery, a ton weight of the material could be produced in half an hour . . . at a cost of less than one shilling per pound."

The target market Parkes was most keenly interested in entering had been described by one of his own associates as "the battle of insulators in telegraphy, which has rivalled that of the gauges in the early days of the railroads." If Parkesine could compete on price with rubber for overhead wires and gutta-percha for submarine wires, Parkesine would triumph in the marketplace.

In April 1866, Parkesine became the toast of Paris (according to Parkes) as England's "most original entry at the Paris Exhibition." It was an ideal time to capitalize on the favorable attention. Parkes accordingly circulated his prospectus for the Parkesine Company, which ringingly repeated his by then well-known claim: "The material can be supplied in any quantity at a cost much below that of India Rubber or Gutta Percha."

But the only way to do accomplish that goal was to resort to the rawest and dirtiest cotton waste to convert into cellulose, as opposed to the cleanest and purest. Rather than systematically control his processes to ensure uniformity of product, he constantly and frantically varied them to meet demand, turning out items of uneven finish. As E. C. Worden sternly castigated him in *Nitrocellulose Industry:*

> Parkes had given out too low costs of production, and in order to substantiate his previous cost estimates, made the material within the price he claimed, but at the expense of quality.

Parkes's "great error," Worden severely concluded, "was in attempting to induce plasticity by adding large quantities of castor oil to the pyroxylin." As a result of this reckless addition of costly, unsuitable solvents, "the combs Parkes sent out in a few weeks became so wrinkled and contorted as to be useless."

Parkes did succeed in turning out a few thousand combs, wrinkled and contorted though they might have been, as well as a few thousand umbrella handles, chessmen, and knife and fork handles. A handful of dentists even tried Parkesine as a substitute for India rubber in making dentures and dental plates. In 1866, the one year that Parkes's factory at Hackney Wick ran at full steam, Parkes even managed to turn out a Parkesine billiard ball, along with some rather attractive ladies' earrings and bracelets, artfully tinted to resemble ivory, coral, amber, and malachite.

What finally did in Parkes and Parkesine was the high cost of raw materials. "From a barrel of mixture," one of his associates ruefully recalled, "the guncotton you got you could hold in your hand." For his part, Parkes blamed these disastrous results on the goadings of money-grubbing associates to abandon "artistic" goals for momentary commercial gain. In his own mind, he was an artist, a bold creator, not an uncouth man of commerce: "When I was manufacturing Parkesine myself my principal object was to make works of art and things of high class value. But when I negotiated this thing on a large scale everybody connected with it urged the necessity of making Parkesine as cheaply as possible—to be dealt in in tons instead of pounds."

Parkes had in essence just summed up the dilemma of the plastics industry for the next, Synthetic Century: balancing the necessity of low-cost manufacturing with the production of high-quality goods. In the eternal battle between art and commerce, plastic was callously shunted off into the column marked "strictly commercial."

After two years of hand-to-mouth operation, Parkes's investors pulled the plug. The Parkesine Company was forced into receivership. The liquidator disposed of the inventory on hand—"1650 dozen white knife handles, 1950 sheets (plain colors and mottled, in a variety of thicknesses), brooches, ear-rings, chains, inkstands, writing cases etc."—at an embarrassingly low price—lower at last than rubber or gutta-percha. In the end, Parkes achieved his elusive goal of undercutting his rivals on price, but at the cost of losing his shirt. From guncotton to Parkesine, those early protoplastics hardly set the world on fire. But there was still a bright future in plastics—you could bet your last nickel on it. Yes, sir.

3

Celluloid Heroes

John Wesley Hyatt
(Courtesy of Hoechst Celanese Corporation)

The ivory brought from the island of Ceylon is the best that can be used for billiard balls, the tusks being far more solid than those from Africa, less friable than those from Continental Asia, and more classic in proportion to their density than any other. They are dreadfully dear, however; and if any inventive genius would discover a substitute for ivory, possessing those qualities which make it valuable to the billiard player, he would make a handsome fortune for himself, and earn our sincerest gratitude. . . .

MICHAEL PHELAN
THE GAME OF BILLIARDS
1859

A Tiffany catalog from the time tells the tale: ivory trays, pin boxes, hat- and hairbrushes, buttonhooks, pincushions, cuticle and steak knives, whisk brooms, and combs; ivory-backed hand and shaving mirrors; ivory-handled shaving, walking-stick, and umbrella handles. Not to neglect those vital necessities of civilized Victorian life, piano keys and billiard balls.

By the mid-1800s, big-game hunters were expressing growing concern that there would soon be no more elephants left to kill. "In part of the northern province of Ceylon," reported the *New York Times* in an urgent dispatch from the killing fields, "upon the reward of a few shillings per head offered by the authorities, 3500 pachyderms were dispatched in less than three years by the natives. Sheffield alone requires annually the slaughter of a large army of the huge pachyderms, estimated some years ago at 22,000, to furnish ivory for the various articles produced in its manufacturing establishments."

The elephant's great enemy was the soaring popularity of ivory knife handles, the best of which were turned out by English factories at Sheffield. Followed by musical instrument keys, mathematical scales, dice, chessmen, billiard balls, and last but not least, "artistic earrings." In the United States, the cessation of hostilities in the Civil War (1865) touched off a sudden upsurge in public interest in sports and games.

A Capital *T* That Rhymes with *P* That Stands for Pool

Within months of the surrender of the South, billiards had blossomed from a prewar fashion into a postwar fad. Pool tables became status symbols in affluent homes, competing with sewing machines and grand pianos—also bad news for elephants. In 1854, America's leading billiards expert, Michael Phelan, revolutionized the "Noble Game" brought back to Europe by the Knights Templar after the First Crusade by inventing the rubber rail cushion. The carom effect instantly transformed a once casual, aristocratic parlor pursuit into a sharper contest of "scientific precision." Five years later, he published a popular guide to the latest postwar innovation, side pockets, which raised the number of balls from three to fifteen, updating snobby "snooker" into the democratic (if sometimes socially unsavory) "pool."

By the mid-1860s, with billiards mania running at fever pitch, Phelan's firm of Phelan & Collender had grown to become America's largest billiards supplier. The billiard-ball business was booming, but a tightening squeeze on top-grade ivory was threatening the very future of the game as a popular pastime. Just one out of every fifty tusks gathered by hunters and poachers in the field was clear grained enough to make a billiard ball. To maintain proper balance, the ball had to be cut from the center of the tusk, steeped in water, and seasoned for up to two years before carving. With the best ivory from Ceylon becoming, as Phelan put it in his billiards bible, "dreadfully dear," his firm took the bold step of running newspaper advertisements around the country, offering a "handsome fortune"—$10,000 in gold—to any "inventive genius" who created a substitute.

In Albany, New York, a twenty-eight-year-old journeyman printer and blacksmith's son from the small upstate village of Starkey spotted one of these newspaper advertisements and eagerly rose to the bait. After leaving home at sixteen to apprentice himself in the printing trade in Illinois, John Wesley Hyatt had experienced some early stirrings of "inventive genius" when he precociously patented—at eighteen—a "family knife sharpener" that operated with emery wheels molded from emery dust fused into a solid mass with a resinous binder. After paying his dues in Illinois, Hyatt returned home to set up a print shop in Albany.

While working full-time at the print shop, he began spending an increasing amount of his spare time—weekday nights and Sunday afternoons—shut up in a small shack behind his house, experimenting with different methods of making billiard balls by compressing odd bits of cloth, wood, and paper into spheres. For binders, he used the glues, shellacs, varnishes, resins, and other adhesives available on the open market. None of these agglutinations quite fit the bill. But by 1865, Hyatt was ready to file for a U.S. patent (50,539) on an imitation ivory billiard ball molded from "layers of muslin or linen cloth coated with shellac and ivory dust" processed under great pressure and high heat.

Hyatt's first billiard balls failed to attain that resonant "click" real ivory balls make when caroming off each other on the green baize. Still, after three years of experimental perseverance, he obtained a second patent describing a more promising mixture of paper and wood pulp, combined with the same weight of shellac, fused together under high heat and pressure.

Hyatt would probably have kept molding this motley collection of pulps, binders, fillers, and glues for years if he had not suffered—like Schönbein, Goodyear, and Perkin before him—one of those lucky spills that often sparked major breakthroughs in the early days of industrial chemistry. Prelinotype printers using pivotal presses employed poorly paid "print catchers" who pulled hot lead from the die at skin-scorching temperatures, and wore what were known to the trade as "Scotsman's gloves"—the bare skin of their hands—considered cheap enough for a Scotsman because "God grew it back." To protect their fingertips from routine abrasion, printers coated them with "liquid cuticle"—printer's jargon for collodion.

In late 1868, after opening the cupboard where the cuticle was kept, Hyatt was shocked to find the bottle tipped over, and some of its contents spilled out across the rough wooden shelf. The dried collodion had separated from the wood and over time congealed into a length of transparent film about the size and thickness, as Hyatt later recalled, of his thumbnail. Gazing down at this sturdy slice of hardened film, it occurred to Hyatt that a billiard ball

coated with such a compound might display just the elusive properties he was looking for. Hyatt was soon so utterly convinced of collodion's potential as a billiard-ball coating that he persuaded his
older brother, Isaiah Smith Hyatt (still working as a newspaper editor in Illinois) to join him in Albany to exploit this promising material's prizewinning potential.

Within a year of his lucky spill, John Wesley and Isaiah Smith
Hyatt were busy dipping spheres of pressed wood pulp and bone
dust into commercial solutions of collodion, none of which seemed
suited (as Parkes had discovered before them) to making "solid articles." Hyatt learned to mix his own "thick paste of soluble cotton"
in a closed mill, which he would then uncover, permitting the paste
to evaporate into a dough, which he "forced accurately" around a
hard wooden core. On April 6, 1869, Hyatt received U.S. patent
88,634 for "an improved method of coating billiard balls by dipping them in a solution of colored collodion."

Hyatt sent a few samples of his collodion-coated billiard balls
to a select group of pool-parlor proprietors, asking them to conduct
a preliminary round of informal field tests. Since the highly nitrated
collodion coating was chemically related to volatile guncotton,
Hyatt wryly recalled that "a lighted cigar, if applied to the ball,
would at once result in a serious flame." Even more alarmingly,
"any violent contact between the balls," a common enough occur-
rence during pool, invariably set off "a mild explosion, like a per-
cussion guncap." One of Hyatt's first testers, a saloon keeper in
Colorado, wrote back to Hyatt that "he didn't mind, but every time
the balls collided, every man in the room pulled a gun."

Hyatt's collodion-coated billiard balls were never produced
commercially. Nor did he ever submit them to Phelan & Collender
to qualify for the gold prize. Although he did persevere "well into
his seventies . . . in his effort to further perfect billiard balls,"
according to Leo Baekeland (of Bakelite fame), Hyatt's lifelong failure to invent the perfect synthetic billiard ball led him straight to a
success far outstripping his original goal.

Plastic Mass

What Hyatt and his brother did succeed in producing—on a set of
steel rollers once used by their blacksmith father for rolling out iron

wagon-wheel rims—was "a transparent slab, one-fourth of an inch thick, fine, and hard as a piece of wood." This slab differed significantly from similar articles prepared by Parkes because its mass production would become commercially viable.

The key difference between Parkesine and Hyatt's compound was that Hyatt had discovered precisely what Parkes had sought but never found: a suitable solvent for rendering pyroxylin (nitrocellulose) into a malleable mass. To Parkes's credit, the right substance had been staring him in the face all along. John Wesley Hyatt, in fact, generously thanked his brother Isaiah for steering him toward some old Parkes patents, in which it was "clearly stated that a little camphor added to the liquid solvent might be beneficial." Seizing on camphor as possibly the elusive solvent for turning nitrocellulose paste into a plastic mass, Hyatt "conceived the idea that it might be possible to mix mechanically while wet solvents with the pulp and coloring matter, then absorb the moisture with blotting papers under pressure, before finally submitting the mass to heat and pressure."

Camphor was key. A decade before, Parkes had used *liquid* camphor (a solution of camphor gum in alcohol) but only in conjunction with other solvents. For centuries, camphor had been distilled by the natives of Taiwan from the sap of a local laurel tree, the world's only source of camphor. While westerners used camphor as a liniment and moth deterrent and easterners carved it into cryptic shapes as a mystical charm to ward off evil spirits, Hyatt used it to dissolve hot pyroxylin pulp under heat and pressure to create a tough, whitish mass "the consistency of shoe leather." It could be turned out—depending on the amounts of camphor added to the mixture—hard as horn or bone or soft as raw rubber.

In a moment of rare jubilation, Isaiah Smith Hyatt coined a name for the new material. He called it celluloid, combining the Greek suffix *-oid* (meaning "like" or "akin to") with *cellulose,* the raw vegetable matter from which it was formed.

In failing to make an ideal artificial billiard ball, John Wesley Hyatt had achieved something far more commercially, socially, and culturally critical. He had created the first thermoplastic—a moldable mass formed by heat and pressure into a shape that it retains even after the forces that shaped it have been removed. A thermoplastic, once formed by flow, can be reheated, resoftened, and remolded again, in an infinitely repeatable cycle. Thermoplastics turned out to be ideal raw materials for the coming Machine Age. Unlike wood, metal, or other natural materials that have to be cut, carved, or shaped by a trained hand—and often polished and finished by skilled craftsmen—a thermoplastic can be easily molded by mechanized methods of mass production and unskilled factory labor. The Hyatt brothers, hands-on blacksmith's sons, had just launched phase one of the Plastic Age: moldable, malleable, protean by nature.

Isaiah Smith Hyatt, more of a marketing man than his brother, promptly rushed the first slab of celluloid over to the American Hard Rubber Company, hoping to arouse their interest in backing its commercial production. The company reserved judgment, but was intrigued enough to send its top collodion expert, Columbia professor Charles Seeley—who had manufactured collodion for the Union army during the Civil War—to investigate this intriguing new raw material. Seeley "came to our place in Albany," John Wesley Hyatt later recalled, "and we conducted the whole process for his inspection, very successfully. He remarked that he had come prepared to detect some chicanery, but could see no deception, and expressed himself satisfied."

The American Hard Rubber Company never invested a penny in celluloid. Though Isaiah had fervently hoped that some rubber company might appreciate the finer points of a viable "artificial rubber," the American Hard Rubber Company, a partner in the notorious Rubber Trust, had every reason to discourage the growth of such an obvious rival. Yet in this detested monopoly, Hyatt realized, lay celluloid's golden opportunity. After receiving their first patent on celluloid, the Hyatts decided to take on the powerful Rubber Trust by targeting one of its most lucrative products: dental plate blanks.

Rubber's Rival

To make celluloid dentures, Hyatt invented a "stuffing machine": a cylinder tapered down at one end to a narrow half-inch nozzle, surrounded by an oil-filled, gas-heated jacket. He fitted the opposite end of this cylinder with a piston and a screw, which could be turned with a lever to force "cakes of incipient Celluloid" through the heated nozzle.

Hyatt's "stuffing machine" was the direct ancestor of the modern injection-molder, hailed by most plastic partisans as the most significant invention in the evolution of plastic after the invention of celluloid itself, since it permitted this inchoate mass to be shaped by flow, precisely, at high-speed and low cost. Hyatt later adapted his "stuffing machine" to the manufacture of rods and tubes, and further modified it by adding a table-mounted blade to slice celluloid sheets—a precursor to celluloid film, the one product destined to ensure celluloid a lasting place in the evolution of modern life.

Dental plate blanks were, to be sure, a bizarre start-up product for the first thermoplastic. But the dean of celluloid historians, E. C. Worden, fully credits "the attempts of dentists to shake off the rubber trust with spurring on investigators, and contributing in no small measure to the early success of Celluloid."

While the first plastic owed its great break to the anger of extorted dentists and the frugality of pool players, Hyatt was still a long way from capturing any substantial share of the dental plate market. While rubber dental plates' greatest shortcoming was the unpleasant aftertaste from the sulfur used to cure them, celluloid plates had a similar drawback: They tasted and smelled of camphor. Hyatt tried everything he could think of to eliminate the camphor aftertaste from his dental plates, including "prolonged heating . . . on the supposition that [camphor] imparted a disagreeable paste to the plate"—not to mention the palate. However, "the pronounced taste of camphor, the apprehension of practitioners over the inflammability of the product, and its warping under the influence of heat" all presented, in Worden's view, well-nigh insurmountable obstacles to celluloid's crusade: breaking the viselike lock of the Rubber Trust.

Adding to celluloid's problems was its tendency to warp alarmingly under heat, which in practical terms meant the first sip of hot tea or coffee. This defect, not surprisingly, was played up by the mortally threatened Rubber Trust, which knew far better than the naive Hyatt how to play dirty pool. The trust was also successful in planting a steady stream of articles in dental journals to the effect that celluloid dental plates were colored with cinnabar, or mercury sulfide. This well-timed attack on celluloid's strongest selling point—that it could be colored with aniline dyes to closely simulate the true flesh tone of human gums far better than rubber—success-fully diffused celluloid's potential impact by scaring dentists away, inhibiting them from making the leap from rubber to celluloid. Hyatt vigorously sought to conduct a countercampaign by running full-page ads in professional journals like *Dental Cosmos,* but even such desperate measures failed to help celluloid dentures capture any substantial portion of the market. Even more significantly for the growth of plastic and synthetics in general, the smear campaign conducted by the Rubber Trust set a powerful precedent for the future. Whenever manufacturers of natural materials felt threatened by the plastic explosion, attacking the upstart invader on health grounds made a highly effective rearguard action.

Bright Lights, Big City

Celluloid might have remained an obscure would-be rival to rubber in the dental field, produced in a backwater, if a group of well-heeled New York investors, led by Gen. Marshall Lefferts, had not intervened. One of the leading capitalists of his day, Lefferts had recently placed a big bet on Thomas Edison by bankrolling his lab-oratory. Intrigued by the prospects of using celluloid to make more than just dental plates, Lefferts's syndicate offered Hyatt a substan-tial capital infusion on the condition that he relocate his enterprise closer to "home"—their home, of course.

By the end of 1872, the year-old Celluloid Manufacturing Com-pany overcame the fears of its neighbors-to-be and successfully occupied a four-story building on Ferry Street in downtown Newark, in a densely populated residential section known as "the Ironbound" because it is surrounded by railroad tracks. "The peo-ple of Newark," the company chemist later recalled, "did not, as a rule, take kindly to a material reported as being made of guncotton and camphor."

Their concern, as it turned out, was well founded. At six in the evening on September 8, 1875, the celluloid plant burst into flames, "utterly destroying all our stock and machinery," Hyatt later recalled, "pushing out the front of the building (which was very weak) and severely injuring several of our men." What went down in local history as the Ferry Street Fire managed to level the factory in under two hours.

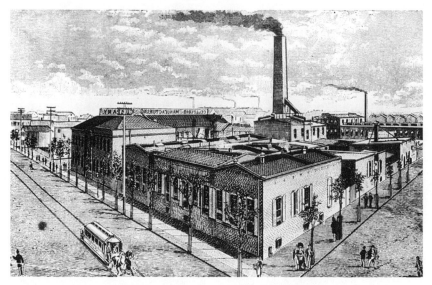

Courtesy of Hoechst Celanese Corporation

The Ferry Street Fire fueled the flames of fear about celluloid's safety at a time when public confidence was desperately needed. It lent new credence to scare stories widely circulated in the popular press, like the one about a formally dressed man at an elegant dinner party who accidentally touched the end of his cigar to one of his celluloid shirt cuffs, setting off a conflagration "from which he and his fellow guests only narrowly escaped with their lives."

Scientific American published a searing report of a young woman who blithely sat down in front of a roaring fire, only to find the celluloid buttons on her dress in flames, forcing her to choose between her modesty and her life.

"The recent explosion in Newark made many people aware for the first time of the existence of Celluloid," reported the *New York Times* the morning after.

> That pleasing compound, which presents itself to the ordinary eye in the guise of a white substance somewhat resembling ivory, is said to be composed of guncotton, camphor and a number of other articles which even the most sanguine and visionary elephant would never dream of attempting to convert into ivory. . . . Nevertheless, after the manufacturer of celluloid

has subjected his guncotton to the action of camphor, the result is an imitation of ivory said to be regarded as a very good imitation by those persons who think it closely resembles ivory. . . .

There could scarcely have been a more qualified endorsement of celluloid's value as an ivory substitute. Of far greater concern to *Times* readers was celluloid's capacity to wreak havoc on its immediate surroundings. With tongue firmly in cheek, the *Times* strove to be reassuring.

Now that a celluloid factory has blown up, it is a matter of great interest to persons possessed of celluloid hairbrushes and billiard balls to know whether these articles will explode. The hairbrush is a favorite tool of ladies who love to drive nails into the wall, and who cannot remember where they last laid away the hammer. But surely a hairbrush likely to explode upon striking a nail and to blow the fair hammerer out the window is to be regarded with grave suspicion. . . .

On an even lighter note—

No man can play billiards with any real satisfaction if he knows that his billiard balls may explode in a series of closely-connected explosions, thereby spoiling a promising run and burying the players under the wreck of the table and cues. Still worse would be the possessor of a set of celluloid teeth who should in a moment of forgetfulness insert the lighted end of his cigar into his mouth.

Celluloid's hazards were mercifully offset by the miraculous ease with which it could be molded and cut, bent, stamped, sliced and otherwise formed into a staggering variety of shapes, textures, and sizes. Some celluloid articles so closely simulated the look and feel of luxury goods fashioned from the world's dwindling supplies of ivory, ebony, horn, tortoiseshell, mother-of-pearl, and bone that consumers were willing to overlook both its real and imagined defects.

In those start-up years, the Hyatts' chief assistants, Charles Burroughs and J. H. Stevens, spent a great deal of time devising clever ways of attaining such shocking levels of verisimilitude. They patented a method of infusing "essence of pearl"—pearl flakes suspended in an alcohol solution—into a celluloid surface, lending it the luster of genuine mother-of-pearl. To simulate the texture and "grain" of bone and ivory, they stacked sheets of celluloid tinted in varying shades of cream, yellow, and white, compressing them under high heat and pressure into a solid block of layered plastic, and then sliced through the striated layers to produce a faux grain indistinguishable by an untrained eye from that of the rare genuine article. Celluloid could also be colored to resemble coral, or marble. Swirled in brown and yellow tints, it resembled tortoiseshell; beige and brown yielded bone. Both tones spawned a new fashion in eye-glass frames.

Personal Effects

Celluloid's great early success came in the fickle field of men's fash-ion, successfully passing itself off as an acceptable alternative (in some quarters) to laboriously laundered fresh "linen"—of the starched collars-and-cuffs variety. All a manufacturer had to do was push hot celluloid against a length of linen to impress the precise texture of the cloth on the white surface. Chemistry had again come to the aid of harried wives and exploited domestics the world over, liberating them from the mundane horrors of constant boiling, bleaching, starching, and ironing.

While some upscale establishments referred to celluloid collars and cuffs by the genteel euphemism of "art"—as in "artificial"—linen, upper-class gentlemen with ample access to domestic service took pride in shunning celluloid collars and cuffs, going so far as to regard them as the latest in an appalling series of vulgar incursions on the traditions of civilized life. When future investment adviser Philip L. Carret graduated from Harvard, class of 1918 (following in the footsteps of his family's first Harvard graduate in 1669), chemistry was his first career choice. But after taking a tour of a grimy gas company laboratory, Carret changed his mind, after observing with fastidious horror his guide sporting a celluloid col-lar, his generation's fashion equivalent of a clip-on tie. Repelled by the sight of "this absolute nadir of sartorial elegance," Carret signed up with the army air force to fly Sopwith Camels over France.

Celluloid may have been a threat to the complacency of "life as we know it," but millions of middle- and clerical-class white-collar workers eagerly snapped up these new washable accessories, advertised with the simple slogan:

> No Laundering Necessary . . . Just a few whisks with a damp
> cloth and collar and cuffs are as clean as new!

Celluloid's appeal to the lower-middle-class clerical worker was reinforced by crude advertising, which depicted weeping Chinese laundrymen wearing silk coats, round hats, and pigtails disembarking in distress from American shores because the all-American celluloid collar had deprived them of work. By so baldly invoking nativist sentiment, celluloid sealed its image as the source of cunning imitations of upper-class luxuries never before available to any but the wealthy. A celluloid collar became a symbol of striving for middle-class respectability, of emphatically not having it as a matter of course.

The same was scarcely true for the distaff side. Throughout the late Victorian era a well-dressed lady would not have been seen in public without "several pounds of celluloid about her person" according to one seasoned social observer. The poundage consisted of clusters of combs to dress flowing locks and tresses. The Celluloid Company thrived on a brisk trade in combs, collars, and cuffs, expanding in Newark to fill a complex of buildings sprawling over thirty acres. Celluloid flourished on faux linen, faux ivory, faux pearl and faux bone until . . .

Spill's Revenge

Out of the blue, disaster struck. In 1877, the company was served with legal papers by one Daniel Spill of Hackney Wick, England, who had traveled to the United States to personally pursue his patent suit. Spill insisted in his formal complaint that celluloid blatantly infringed upon his 1869 English and American patents covering a nearly identical material called Xylonite.

Originally trained as a doctor, Daniel Spill never practiced medicine but instead joined his older brother George in manufacturing rubberized waterproof cloth at a small factory in bleak, industrial east London. The Spills' big break came during the

Crimean War, when the rain-soaked British army placed an order for 340,000 yards of rubberized cloth from a "Mr. George Spill of Stepney Green" for waterproof capes and ground cloths.

The Spill brothers capably rose to this challenge by churning out seven miles of waterproof cloth every day, "which was then cut up into garments, the making of which gave employment to several thousand hands." The successful completion of this large government contract not only endeared the firm to Her Majesty's government, but encouraged them to relocate to larger premises at Hackney Wick, near Victoria Station, from which they mounted an impressive display of "waterproof vegetable fabric, mineralized, of Superfine India Rubber Cloth, weighing just six ounces, warranted to withstand heat under 400 degrees Fahrenheit" at the 1862 Great International Exhibition.

As it so happened, the Spills' fine waterproof cloth was awarded the bronze medal for "versatility of invention" in the same category—"Vegetable Materials Used in Manufacture"—as Parkesine. Daniel Spill, by then George's plant manager, occupied a display booth directly opposite Parkes's. The younger Spill, intrigued by Parkes's claim that his new material was not merely waterproof but could be "spread or worked in a similar manner to India rubber," approached Parkes with a proposal to experiment with Parkesine as a laminate for a new line of waterproof garments. Parkes, "determined by then to manufacture solid articles principally composed of nitrocellulose and to manufacture them at the lowest possible cost," was pleased by the arrangement, because in return for giving them an exclusive on Parkesine as a waterproof coating, the Spills agreed to manufacture Parkesine in bulk at their new works in Hackney Wick.

After just two years of operation, the Parkesine company collapsed. But the resourceful Daniel Spill persuaded a group of investors (including the British representative of the Hyatts' Celluloid Manufacturing Company) to put up £143,000 to reorganize as Daniel Spill & Co. While the Hyatts in Albany were busy coating wooden cores with collodion, Spill filed for and was granted British and American patents in 1869 for a pyroxylin product called Xylonite.

He promoted Xylonite much as Parkes had promoted Parkesine:

As a substitute for Ivory, Bone, Horn, Tortoiseshell,
Hard Woods, Vulcanite, Papier Mache, Marbles, Brass,
and Veneers for Cabinet Work. Can be Also Applied to
Waterproofing Fabrics, Leather, Cloth, Book Binder's Cloth,
Card Cloth, Writing Tablets, Bagatelle Balls, Pianoforte Keys,
Gear and Friction Wheels, and Bearings for Machinery, Spinner's
Bosses, and as an Insulator and Protector of Telegraph Wires.

Unfortunately, Spill failed to learn from Parkes's great mistake: the high cost of the large volumes of solvents needed to produce a plastic mass by Parkes's methods. Spill's processes were marginally more efficient in the use of solvents than Parkes's, but still he consumed a ton of solvents for every fifty pounds of plastic produced. Given such a sorry ratio of solvent to product, there was no way that Xylonite could ever be profitable. By 1875, Spill's new company had followed its predecessor into bankruptcy.

Two years later, remarkably undeterred by this series of setbacks, the intrepid Spill was up and running again, now calling itself The British Xylonite Company Ltd. But this time around, Spill had an insurance policy in mind. Rather than rely entirely on the vicissitudes of the marketplace, he would see John Hyatt in court and win from him a large enough settlement to corner the British market. Legally and scientifically, camphor lay at the crux of the case. In his original patent, Hyatt had specified the use of "comminuted (ground) camphor gum in conjunction with nitrocellulose" as his basic recipe for celluloid. Spill challenged Hyatt's primacy in this department by claiming an earlier use of camphor and alcohol as solvent for pyroxylin as far back as 1869, when he was granted a U.S. patent for the use of camphor and "spirits of wine"—ethyl alcohol—in manufacturing Xylonite. So the key question became whether Hyatt, Parkes, or Spill had been the first to grasp the fundamental value of using camphor, and camphor alone, and in solid not in liquid form, in making a pyroxylin "plastic mass."

Throughout the ensuing titanic legal battle, Alexander Parkes demonstrated his resentment of his former plant manager's temerity in putting himself forward as Britain's leading "plastic man" by doing whatever he could to undermine his old colleague's case. By claiming under oath that he was "the first to discover the fact that camphor alone was a solvent of nitrocellulose," he did what he could to invalidate Spill's claims that he had been first to use it.

That this was patently untrue did not deter Parkes in the slightest. Nevertheless, in May 1889, the Honorable Judge C. J. Blatchford, clearly no chemist, found in favor of Spill. Judge Blatchford not only awarded Spill the then staggering sum of $750,000 in damages but took the radical step of prohibiting Hyatt from using camphor and ethyl alcohol as solvents in the manufacture of celluloid. But rather than closing up shop, Hyatt simply suspended the use of *ethyl* alcohol and substituted *methyl*—wood—alcohol. Upholding the letter if not the spirit of Blatchford's injunction against "spirits of wine" (an archaic term for grain alcohol) kept John Wesley Hyatt in the celluloid business.

An enraged Spill did precisely what Hyatt had hoped he would do: He hauled Hyatt back into court, seeking to enforce Blatchford's injunction. But once on the stand, while preparing to provide Blatchford with his estimation of the damage done him by Hyatt's infringement, Spill was devastated by the presentation by Hyatt's lawyers of new evidence, uncovered during an extensive patent search, which proved beyond a reasonable doubt that it had been Parkes, not Spill, who had been the first to employ "vegetable naphtha" (also known as wood alcohol) as a solvent in Parkesine.

On the basis of this new evidence, Judge Blatchford had little choice but to reverse his earlier judgment. But once again, he missed the point entirely by ruling that Parkes had been the first to use camphor and camphor alone to make a pyroxylin product, thereby invalidating Spill's and Hyatt's claims to primacy in the matter. At the stroke of a pen, Blatchford had thrown open the celluloid market—and Hyatt's patented processes—to imitation by anyone with the will and the means to do so.

First out of the gate in competing with celluloid on the American market was none other than Daniel Spill, whose American Zylonite Company was soon churning out tons of celluloid-like articles—combs, belt buckles, brush backs, and the like—at a small plant in North Adams, Massachusetts. Still, Blatchford's decision to invalidate his patent claim was a momentous setback for Spill, from which he never fully recovered. Two years after his return to England in 1885, he died of diabetes, fifty-five years old, a broken, embittered man.

Photo Finish

In those days one did not 'take' a camera. One accompanied the outfit of which the camera was merely a part.

—George Eastman

In 1874, twenty-year-old George Eastman was a poorly paid clerk in Rochester, New York, working long hours for an insurance company located in the same office building that had once housed "Eastman's Commercial College," a business school founded by his father. After George Washington Eastman's premature death in 1862, the college was forced to close its doors. The Eastmans were plunged into abject, if genteel, poverty—Mrs. Eastman was obliged to take in lodgers. At fourteen, George left school to help support his sainted mother.

After serving six years as messenger for the insurance company at three dollars a week (while studying accounting in the evenings), Eastman got his first break: the Rochester Savings Bank offered him a job as a junior clerk at the then magnificent salary of $800 per year. On that sort of pay, a young fellow who had been living at home all those years, scrimping and saving to help his mother pay her monthly bills, could afford a few petty indulgences. But Eastman waited another three years, until pulling in the princely sum of $1,400, before taking his first tentative plunge into high living.

The frugal Eastman decided to take a trip to exotic Santo Domingo. Being a puritan at heart, he felt obliged to make a complete visual record of his first trip abroad so that the momentous experience would not simply "go to waste." As a purely practical matter, Eastman signed up for photography lessons (at $5 an hour) with a local portrait photographer, George Monroe. Monroe sold him a full photographic outfit, including "lenses and other sundries," for just under $100.

The sundries presented a problem. As Eastman later recalled:

My layout, which included only the essentials, had in it a camera about the size of a soap box, a tripod strong and heavy enough to support a bungalow, a big plate-holder, a dark-tent, and a big container of water. Since I took my views mostly outdoors—I had no studio—the bulk of the paraphernalia worried me. It seemed that one ought to be able to carry less than a pack-horse load to take pictures.

Even more humiliating than service as a human packhorse was the absurdity of toting all this heavy stuff around in a "photographic perambulator"—a wheelbarrow that held the collapsible "dark tent," which folded into a wooden case the size of a large suitcase—along with a motley assortment of glass bottles required for processing wet collodion plates on the spot.

As a dress rehearsal for Santo Domingo, Eastman took a short jaunt to Michigan's rugged Mackinaw Island. His photo safari went off without incident except for one minor mishap: When he reached into his kit, he discovered that a glass bottle of silver nitrate he had wrapped in his underwear to keep from breaking had broken—and leaked all over the contents of his bag. Having to replace a perfectly good pair of underdrawers compelled Eastman into moody contemplation of glass's manifold drawbacks. It was heavy, clumsy, breakable.

Meanwhile, Back in Rochester

On a cold morning in January 1878, while sitting in his mother's kitchen by a hot stove over breakfast, Eastman (by then twenty-four) happened to glance at a glowing account in the *British Journal of Photography* by the British photographer Charles Bennet of a new "heat-ripening" process by which he had dramatically increased collodion's light sensitivity by heating the emulsion on a hot stove.

For decades since the appearance of the Archergram, desperate photographers who wanted to capture images outdoors tried everything they could think of to retard collodion's rapid rate of evaporation. Since the photograph had to be developed before the collodion was dried—the reason it was known as a "wet" process—the picture had to be taken and an elaborate series of procedures performed in complete darkness, in hot, unventilated "dark tents" in a matter of seconds, not minutes.

Photographers tried adding sugar, beer, tobacco, and licorice to collodion. They even spat on it in a pinch. But nothing helped until 1871, when an English physician, Dr. Richard Leach Maddox—who despised wet plates because the evaporating ether fumes damaged his lungs—tried coating glass plates with a gelatin emulsion.

Maddox's "dry plates" were a revolution: The gelatin coating slowed down the evaporation of the collodion, dramatically expanding the possibilities of outdoor photography. Five years later, another amateur British photographer, Richard Kennet, further improved Maddox's process by developing packets of dried gelatin emulsion—he called them "pellicles"—which could be prepared with the ease of fixing a cup of tea. And two years after that (1878) yet another British amateur, Charles Bennet, discovered that heating the gelatin pellicles would "ripen" collodion, boosting its light sensitivity.

What if, young Eastman wondered, *he* could employ Bennet's "ripening" process to prepare dry plates in volume and supply them to local photographers? Eastman reckoned that preprepared dry plates would save them untold time and effort, dramatically lighten their already crushing load, and—no small concern—possibly free him from his desk job at the bank. However, Eastman's prime motivation, he later allowed, was to support his mother in the comfortable style to which she had once been accustomed.

That night, Eastman set up a small photographic laboratory, appropriately located in his mother's kitchen, and after turning in a full day's clerk work at the bank, stayed up until the wee hours of the night cooking up his emulsions. Some nights, he was so spent by his labors that he fell asleep in his clothes on the kitchen floor, spreading himself out near the still warm stove to keep from catching cold.

By July 1879, though still employed full-time at the bank, Eastman had achieved impressive results. He had perfected a plate-coating machine that pulled a glass plate across a rubber roller partly immersed in a trough of melted emulsion. He was thus able to mechanically produce a thin, uniform gelatin coating, thereby eliminating at the turn of a crank the need to spread gelatin by hand with a honey stick.

Drawing on some prudent portion of his meager savings, Eastman took his first trip abroad (he had, characteristically, abandoned his trip to Santo Domingo, citing lack of funds as the cause). Booking passage to England, he hoped to obtain a British patent on his plate-coating machine, the rights to which he would then sell to raise funds for his Rochester start-up. With the then handsome sum of £500 (worth $2,500 in the nineteenth century) in his pocket, obtained from selling the English patent to his invention, he signed a lease on a loft on State Street in downtown Rochester, but continued to operate the company as a sideline while working full-time at the bank.

On January 1, 1881, Eastman formally incorporated as the Eastman Dry Plate Company, having by then acquired a financial backer and partner, a whip-and-buggy manufacturer named Henry A. Strong, one of his mother's boarders. Not until revenues topped $4,000 in cash every month did he feel able to quit his day job, but after leaving the bank, the company nearly collapsed when a large number of dry plates turned out to be defective.

A large backup of orders had forced him to stockpile hundreds of precoated dry plates, which he had kept in storage until he was ready to ship them to customers. As it turned out, sitting in storage caused them to lose their light sensitivity on the shelf, a fact that only emerged after a posse of irate photographers found that the plates would not produce images. A financially strapped Eastman, eager to restore his damaged reputation, was forced to dig into his shallow pockets to make good on the bad plates. And in the critical first few months of operation, he was forced to take the time out to travel to Britain to seek advice on what had gone wrong.

As it turned out, the rotten apple that had killed the whole barrel turned out to be nothing more than a bad batch of gelatin. But the seriousness of the incident compelled him again to consider the prospect of dispensing with glass. He found a short solution to his problem in an 1881 review in *Popular Science* of his own "Progress in Photography." While conceding that Eastman's precoated dry plates offered "considerable advantages," the author confidently predicted the ultimate development of a flexible "film" that could be rolled up inside the camera. Even the term itself was brand new in this usage, derived from the Anglo-Saxon word *filmen,* meaning "the scum that forms on boiling milk."

Eastman was not the first to experiment with plastic film for photography. The notion by then had a short if frustrating history. Alexander Parkes had prepared the first thick collodion film for photography, hoping to find a substitute for collodion that would be stout enough to dispense with the underlying glass support. In 1870, Daniel Spill had addressed the prestigious London Photographic Society on the "future of a flexible and structureless substitute for glass negative supports," which he had obviously hoped would be Xylonite. In 1875, an English photographic supplier, Leon Warnerke, succeeded in perfecting a roll-film system that relied on "stripping film"—paper coated on one side with a light-sensitive collodion coating. But Warnerke's "stripping film" had to be stripped from its paper base before being developed, a delicate task that required the touch of a surgeon combined with the patience of a saint.

Although Warnerke's film holder was too costly and cumbersome to be commercially viable, it provided Eastman with a fine working model. A born mechanical genius, Eastman's ability was firmly grounded in an acute sensitivity to his own limitations. He had a knack for recruiting people to fill his personal voids. Having recently invented the best—indeed the only—plate-coating machine in the field, Eastman saw no reason why he shouldn't be the one to perfect Warnerke's system. But he intuitively knew he needed a partner, one blessed with mechanical ability and considerable photographic expertise.

He found just the man he was looking for in January 1884—a Rochester camera manufacturer, William Walker, who had recently put on the market a highly sophisticated folding camera that mightily impressed Eastman with its technical finesse. Walker agreed to close down his own shop in exchange for the chance to go in with Eastman on a new venture: the development of the first commercially viable roll-film system.

By the end of the year, Eastman and Walker were ready to roll. Like Warnerke's, their system relied on paper "stripping" film generously coated with gelatin. But compared to Warnerke's costly and cumbersome mahogany frame, the Eastman-Walker roll-film holder was a frugal device, reflecting Eastman's acutely cost-conscious personality. It conveniently fit into any standard glass-plate camera, and was capable of holding enough film for twenty-four exposures.

Any resemblance between his roll-film holder and Warnerke's, Eastman stoutly insisted, was pure coincidence. As he thundered at one probing reporter, "There is about as much similarity between it and our holder as there is between an old flint-lock blunderbuss and a Smith & Wesson self-acting six-shooter!"

Derivative or not, a sure sign of Eastman's unflagging faith in the future of flexible film was that immediately after introducing the roll-film holder he renamed his company The Eastman Dry

Plate & *Film* Company. He followed up by hitting the market with a superior emulsion called American film, in which the light-sensitive gelatin emulsion was separated from its paper base by a thin layer of water-soluble gelatin. The image-bearing light-sensitive gelatin emulsion still had to be detached from the paper and transferred to a sheet of clear gelatin, then coated with collodion before processing, but it did, as Eastman advertised, "combine the lightness and convenience of paper with the transparency of glass." Still, Eastman knew that to accomplish his ultimate goal of replacing glass altogether he needed to come up with something better: a "plastic" film strong enough and tough enough to support the emulsion on its own, without the aid of a paper backing.

Just as he had sought out William Walker for help on the mechanical side of the roll-film dilemma, Eastman knew that he needed a first-rate chemist to improve the film itself. Fortunately, he found one at the new chemistry lab at the University of Rochester, which had been funded by one of Eastman's old colleagues at the Rochester Savings Bank, and was headed by a Professor Samuel Lattimore. Lattimore's graduate student assistant, Henry M. Reichenbach, struck Eastman upon first meeting as "an ingenious, quick-witted fellow." Eastman made him an offer on the spot. "He knows nothing about photography," the young entrepreneur cheerfully told a skeptical Walker. "But I told him what was wanted and that it might take a day, a week, a month or a year to get it, perhaps longer, but that it was a dead sure thing in the end." The dead sure thing that Eastman wanted was a practical, flexible film.

As for raw materials, one obvious candidate presented itself. "Collodion had always fascinated Eastman," Eastman's biographer Carl Ackerman observed. "Here was a substance made from cotton and nitric acid. . . . The fact that such a chemical union produced both collodion and an explosive had natural attractions for a man who had experimented as much as Eastman."

Eastman's main stumbling block was that he had been "unable with one coating to get a sufficiently heavy skin or Pellicle of collodion to serve as the final support for the emulsion." This was precisely the job Eastman hired Reichenbach to handle for him. With Reichenbach working full-time on the film side, Eastman felt free to fix his sights on another pet project: a compact camera that would accept the new flexible film, be simple to operate, and cheap enough for Everyman.

In June 1888, Eastman rolled out the first of his plain wooden boxes that would catapult expensive, aristocratic photography squarely into the Age of Democracy. It was called a "Kodak," he later insisted, because "this purely arbitrary combination of letters . . . displayed a vigorous and distinctive personality." He neglected to mention that the letter *K* was also the first letter of his mother's maiden name.

The Kodak weighed a pound and a half and cost Eastman just $6.35 to make. It cost the customer $25.00, including a "hand-sewed leather Carrying Case and shoulder strap" and a roll of American film long enough to make a hundred two-and-a-half-inch-diameter circular exposures.

Before Kodak, photographs had to be developed by the photographer himself, or by professional portrait studios, which charged exorbitant prices. With Kodak, all a customer had to do once the roll of a hundred pictures had been exposed was to send the camera back to Eastman for processing. Eastman developed the film, printed it, and returned the camera to the customer, all for a flat fee of $10.

"The principle of the Kodak system," proclaimed a primer provided to every proud owner of the new camera, was

> the separation of the work that any person can do . . . from work that only an expert can do.
>
> *Ten years ago* . . . every photographer had to sensitize his own plates and develop and finish his own negatives. This necessitated the carrying of a kit and paraphernalia of a studio with him into the field. . . .
>
> *Four years ago*, the amateur photographer was confined to heavy glass plates for making his negatives, and the number of pictures he could make was limited by his capacity as a pack-horse. . . .
>
> *Yesterday*, the photographer, whether he used glass plates or films must have a darkroom and know about focussing, developing, fixing, intensifying and mounting before he could show good results from his labors. . . .
>
> *Today* . . . photography has been Reduced To Three Motions: 1) Pull the Cord 2) Turn the Key 3) Press the Button.

All this would soon be reduced to a simple slogan: "You Push the Button, We Do the Rest."

Back in his lab at Rochester, Henry Reichenbach was making steady progress on his secret assignment. His first breakthrough came when he discovered that in 1882 John Henry Stevens of the Celluloid Company (the kid who had been hired at fourteen by the Hyatts as an office boy) had perfected a process by which amyl acetate (banana oil) could be added to celluloid to improve its flexibility. Stevens's goal had been to produce celluloid in the form of a liquid varnish. Reichenbach already understood that a film is nothing more than a dried varnish, and his first nitrocellulose preparations relied heavily on Stevens's amyl acetate recipe.

Those first solutions flowed easily onto a glass plate and dried evenly, without bubbles or flaws, producing a smooth, clear, brittle film whose only significant drawbacks were a tendency to tear easily and a nasty habit of peeling away from the glass after drying. It took Reichenbach months and many hundreds of experiments—he tried adding camphor to increase the film's tensile strength and flexibility, but it turned the film cloudy—before he found the solution. The key turned out to be fusel oil, a by-product of whisky distilling.

For this loyal service, his boss was properly grateful. When the time came to apply for a patent in March 1889, Eastman magnanimously insisted that "Reichenbach's name be connected with all patent applications—I think it would please him." The boss even took such uncommon generosity one step further by giving Reichenbach full credit for "the chemical part" of inventing celluloid film. As for "the mechanical part," that was the work of George himself.

The "mechanical part" of the film-casting process was impressive. Eastman's manufacturing line took up a room twice the length of a football field, with a vaulted ceiling from which was suspended on steel chains a large hopper filled with nitrocellulose "dope." The hopper glided along steel tracks along the ceiling at a steady rate, pouring precisely calibrated amounts of dope onto rows of highly polished, two-hundred-foot long, three-and-a-half-foot-wide plate-glass tables, finely ground at the joints where the sheets of glass met. The thin liquid sheet thus produced was allowed to set overnight until dry. The next morning, the transparent film was stripped from the tables with spatulas, rolled and slit down the middle, and finally wound onto spools.

Writing to a colleague, Eastman could scarcely contain his enthusiasm for this latest sparkling invention: "This new film is the 'slickest' product we've ever made. The field for it is immense!" He knew that his flexible film would not only revolutionize photography but would permit the development of new cameras that would in time turn an aristocratic art form into a truly mass medium.

In June 1889, Eastman launched a major marketing blitz for transparent film by sending out free samples to friendly editors of the leading photographic trade journals. Photography was still a gentleman's hobby, patronized by a tightly knit elite dominated by affluent amateurs. The new film had its first public demonstration at an August meeting of New York's Society of Amateur Photographers, where Eastman representative G. D. Milburn gave a stirring presentation hailing his boss as "the inventor of flexible-film photography."

A gentleman sitting at the back of the room begged respectfully to differ. John Carbutt, a prominent Philadelphia photographic supplier, took this opportunity to publicly put forward a rival claim for the true "inventor" of celluloid film, on behalf not of himself but of an as yet unnamed individual.

> Some three or four years ago my investigations in celluloid commenced. And since I have become a manufacturer of the article I have learned that I was antedated by one or two years. So I have to disagree with the gentleman who preceded me in the statement that Mr. Eastman was the first to produce transparent film. I was shown, in New York . . . a film said to have been made five years ago on celluloid. I have no reason to dispute that as being a fact. . . .

Eastman's claim to having pioneered celluloid film was in the end challenged not only by Carbutt but by another contender, the as-yet anonymous "antecedent" of Carbutt's cryptic comment. Eastman's secret rival turned out to be an obscure Episcopal priest, the Reverend Hannibal Goodwin, a rank amateur who in the early 1880s had stumbled upon celluloid as a possible substitute for glass plates for illustrating his Tuesday-night lectures on biblical subjects at his rectory in Newark, New Jersey.

Out of sheer frustration at being able to purchase glass slides of Niagara Falls and Yellowstone Park, of the Capitol in Washington, and the newly built elevated railway in New York City, but none of Jerusalem or Bethlehem, Goodwin approached the Celluloid Company—his church was a neighbor—for help in replacing clumsy, breakable glass plates with a celluloid film base. Despite a lack of funds for research and a complete ignorance of scientific principle, training, or theory, Goodwin against all odds actually succeeded in perfecting a viable celluloid film. He filed for a prospective patent for his "Photographic Pellicle and Process of Producing Same," on May 2, 1887—two years before Eastman and Reichenbach's similar patent application.

While the Eastman-Reichenbach joint application was quickly approved in April 1889, only a month after it was filed, a different patent examiner didn't issue Goodwin his U.S. patent 610,861 until September 1898. The patent examiner, for no easily explicable reason, delayed processing Goodwin's application for *eleven years*. No evidence of outright chicanery or conspiracy ever surfaced to explain Goodwin's kafkaesque trials at the hands of the United States Patent Office. But there was little disputing that Goodwin had been first to file, if not necessarily to be granted a patent, for a flexible celluloid film. Carbutt, for his part, never bothered to patent his celluloid sheet film, which was much stiffer than Eastman's and was never intended to be rolled up inside a camera.

Regardless of the validity of Goodwin's claim, George Eastman was not about to give up the battle without a fight, a nasty one if need be. As the deeper-pocketed defendant he knew the soundest strategy would be to employ every legal instrument at his command to drag out the ensuing suit for as long as possible, a goal he succeeded admirably in attaining, by delaying the verdict for twenty-five years, until 1913, by which time the reverend was long dead. But Goodwin, posthumously at least, was destined to have the last laugh. After selling his patent rights to the New York photographic supply house of T. H. Anthony & Company for a handsome sum, mainly on the strength of his suit, Goodwin's heirs considered themselves well compensated when Eastman was obliged by the judge to pay $5 million in belated recognition of their ancestor's astounding achievement.

Celluloid film's resounding success—by the turn of the century the market was estimated at eighty million pounds per year—set plastic free to stand on its own. Due to Judge Blatchford's erroneous ruling, John Wesley Hyatt never derived a penny of profit from the voluminous sale of celluloid film. No longer just a cheap imitation of natural materials but a vital medium of visual information storage, celluloid film succeeded in raising the first plastic's cultural profile from a medium of mere mimicry into a priceless repository of human memory. Or, as Eastman himself put it in one of his pompous pronouncements on the subject:

> Photography is brought within the reach of every human being who desires to preserve a record of what he sees. Such a photographic notebook is an enduring record of many things seen only once in a lifetime. . . .

That lifetime ended in 1930, twenty-three years after his mother's death.

George had continued to inhabit the thirty-seven-room Georgian mansion he built for her in Rochester, with its many unoccupied bedrooms carefully labeled alphabetically *A, B, C, D,* on up to the key letter *K.*

At seventy-six, the lifelong bachelor was diagnosed as suffering from a progressive spinal disorder. Two years later, after smoking a last Lucky Strike—his favorite brand—Eastman stubbed out his cigarette, took off his glasses and skullcap, and lay down on his massive bed. After carefully placing a short note, written on a sheet of lined yellow foolscap, on a bedside table beside his mother's sewing basket, he wrapped a wet towel over his chest (to prevent powder burns) and shot himself in the heart with a Luger automatic pistol.

To my friends. My work is done. Why wait? GE

Bodies in Motion

> The great and lasting contribution of Edison [to the development of cinematography] was his use of celluloid film, 35mm. wide, with four perforations for each picture, a practice still standard sixty years later.
> A HISTORY OF TECHNOLOGY OF THE LATE NINETEENTH CENTURY (1850–1890)

In 1887, Thomas Alva Edison, inventor of the incandescent bulb, father of the phonograph, founder of electricity, developer of the multiplex telegraph, transferred the bulk of his manufacturing operations from New York City upstate to Schenectady, as a way of "getting away from the embarrassment of the strikes and the communists to a place where our men are settled in their own homes."

In a similarly escapist mood, he purchased fourteen acres in Orange, New Jersey, the largest of four then sylvan suburbs of Newark known as the Oranges. There, after the fashion of an industrial Kubla Khan, he decreed the construction of a massive, 250-foot-long, 100-foot-wide brick building described by one awestruck visitor as a "huge but not ungainly pile of red brick with its succession of wings thrown out like the teeth of some Brobdingnagian comb."

In the midst of pitched battles with his arch nemesis George Westinghouse over the distribution of electrical power, and with the financier Jesse Lippincott over a fraudulent scheme to deprive him of his rights to the phonograph, Edison moved into his spacious new laboratory. Once ensconced, Edison later insisted that an incandescent light bulb went off in his head.

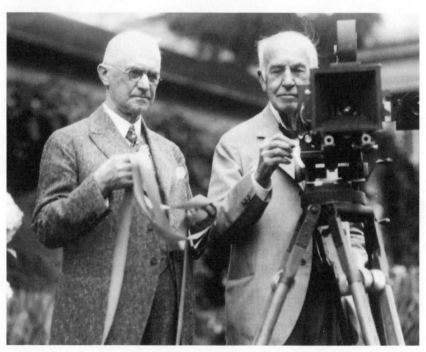

Thomas Edison and George Eastman
(Reprinted with permission from The Eastman Kodak Company)

The idea occurred to me that it was possible to devise an instrument which should do for the Eye what the phonograph does for the Ear, and that by a combination of the two, all motion and sound could be recorded and reproduced simultaneously. . . .

The reality, of course, was not quite so simple. If the truth be told—and in Edison's case, it frequently wasn't—it probably didn't occur to Edison to propose making motion pictures until Saturday night, February 25, 1888, when the British photographer Eadweard Muybridge came to Orange to give a standing-room-only lecture sponsored by the New England Society.

All Orange society was thrilled to be entertained by the celebrated photographer whose "magic lantern show" of animals and people in motion had become hugely popular as a thinking man's vaudeville. Muybridge—a native of Kingston-on-Thames who had lived happily with the more pedestrian name Edward James Muggeridge until adopting the poetic Anglo-Saxon rendition—emigrated to America at thirty and set himself up as a bookseller in San Francisco. After sustaining a powerful blow to the head during a serious stagecoach accident, he had been urged by the doctor who treated him for "sensory deprivation" to take up a career that would oblige him to spend more time outdoors.

He wrangled a job as a survey photographer with the War and the Treasury departments during the Civil War, and acquitted himself honorably enough to be nominated official photographer for the first government expedition sent out to survey the nation's new national parks.

Muybridge's stunning views of Yosemite attracted huge crowds to his exhibitions, which brought him to the attention of Leland Stanford, the former governor of California and founder of Stanford University, who in 1874 hired him to settle a bet.

Stanford owned a string of champion racehorses, of which his trotter Occident was by far his favorite. While singing the praises of Occident's pure poetry in motion, Stanford insisted one night at his club that at some point during his trot down the track, the horse must—for a split second at least—lift all four legs "simultaneously free from contact with the ground," as Muybridge later put it. Stanford's fellow horse lover Frederick MacCrellish was only too happy to bet Stanford $25,000 that this proposition of "unsupported transport" was bunk.

Inspired by the work of Étienne-Jules Marey, the French physiologist whose photographic documentation of animal locomotion—"the horse's gallop, the bird's flight, the insect's quiver"—were the scientific sensation of his day, Muybridge developed a fast shutter capable of making exposures of one five-hundredths of a second. Having made a name for himself in landscapes, he was determined to capture bodies in motion.

Stanford approached Muybridge with a proposition: How would he like to test his new stop-action shutter, sparing no expense of course, to document Occident's trot right down to the last hoofbeat? Intrigued as Muybridge was by this proposal, as he was about to stand trial in San Francisco for the murder of his wife's lover, he was forced to put Stanford off. Though acquitted on a technicality, Muybridge fled the country soon after the verdict was announced, not to return for four years, until the public furor had died down.

In 1877 (the same year Daniel Spill sued John Wesley Hyatt and George Eastman took up photography) Muybridge established himself at Stanford's palatial stud farm in Palo Alto—later the site of Stanford University—and erected an elaborate apparatus consisting of twenty-four cameras, twelve to a side, set up along a specially prepared racetrack covered in corrugated rubber so that no dust and cinders would be kicked up by the horse's hooves to obscure the clarity of the images. Each camera was outfitted with an electromagnetic shutter, linked to a trip wire, stretched taut across the track to a white-painted screen squared with black lines, resembling a piece of graph paper. As Occident raced down the track, the wheels of the sulky tripped the guy wires, releasing the shutters.

As Muybridge later observed:

> We made several negatives of a celebrated horse named Occident while trotting laterally in front of a camera—the resulting photographs were sufficiently sharp to give a recognizable silhouette portrait of the driver, while some exhibited the horse with all four of his feet clearly lifted, at the same time, above the surface of the ground.

Not only did Stanford win the $25,000, but Muybridge embarked on a startling new career as an animal dynamics and locomotion expert. But the most sensational aspect of Muybridge's work was yet to come. By pasting the opposite halves of a series of stereoscopic pictures into a pair of *zoetropes*—a children's toy made from an open drum repeatedly slit on the side—and spinning the pair of *zoetropes* before a light source, "a very satisfactory reproduction of an apparently miniature horse trotting . . . could be obtained."

During his well-documented two-hour visit with Thomas Edison on the Monday following his lecture to the New England Society, the two men discussed—according to Muybridge—the possibility of combining Muybridge's *zoopraxiscope* with Edison's phonograph. But the duplicitous Edison, seeking to cover his tracks later on, went so far as to scrawl in the margin of a manuscript of an early biography of himself, correcting a common impression: "Muybridge came to Lab to show me a picture of a horse in motion—*nothing was said about phonograph.*"

Nonetheless, a few months after that encounter, Edison set down a few paragraphs in a "caveat"—a term used in those days to describe an as-yet-to-be patented invention—in which he claimed to be

experimenting upon an instrument which does for the Eye what the phonograph does for the Ear, which is the recording and reproducing of things in motion, in such a form as to be Cheap, practical and convenient. This apparatus I call a Kinetoscope "Moving View." In the first production of the actual motions—that is to say of a continuous Opera—the Instrument may be called a Kinetograph . . . but is properly called a Kinetoscope.

William Laurie Kennedy Dickson, who had emigrated from Scotland to America in his teens with the sole ambition of working with Edison, had by 1884 landed a job at Edison's New York factory. By the following year he was being referred to around the premises as Edison's "official photographer." Five years later, Edison gave Dickson a brief break from his other duties to begin conducting his now famous "Magic Lantern" experiments.

Dickson commenced by taking an aluminum drum slit down the sides, comparable to Muybridge's *zoopraxiscope,* and covering it with some two hundred "pin-point microphotos" arranged in a spiral. He hoped to synchronize the turning of the drum in front of a light source with a spinning phonograph cylinder. In his first "Motion Picture Caveat," Edison wrote: "The cylinder may even be covered with a shell and a thin flat film or transparent tissue light-sensitized."

The "thin flat film" or "transparent tissue" Dickson employed was John Carbutt's stiff celluloid sheeting, made expressly for Carbutt by John Stevens of Hyatt's Celluloid Company in thicknesses of one two-hundredth of an inch. But he was soon writing imploringly to the American Zylonite Company (the American subsidiary of Spill's British Xylonite Company) in search of a more flexible transparent film, which could be more easily wrapped around a photocylinder. Zylonite responded by offering to bend its celluloid sheets "into any shape desired," but even that was no dice: No film proved flexible enough. Until, a few months later, Dickson attended the meeting of the Society of Amateur Photographers in New York where John Carbutt first challenged George Eastman over the invention of celluloid film. Dickson could not have been less concerned about whether Carbutt, Eastman, or Goodwin had been first to invent it. He was strictly concerned with the film's flexibility, and after Eastman representative G. D. Milburn concluded his presentation, Dickson persuaded Milburn to part with his only sample—so it could be shown directly to Mr. Edison.

According to Dickson's subsequent account, that night Edison took one look at Eastman's new transparent film and exclaimed "with a seraphic smile": "That's it—we've got it. We can now do the trick—just work like hell!"

Which makes a colorful story, except for one minor detail: On the date in question, Edison was in Paris attending the 1889 French Exposition, holding long fruitful talks with Étienne-Jules Marey, who had been hard at work on his own motion-picture apparatus, also using strips of the new Eastman flexible film.

Marey had taken his "chronophotographs" with a photographic "revolver," an invention inspired by an earlier conversation with Muybridge in Paris. After inspecting Muybridge's *zoopraxiscope*, Marey's equipped a "rifle" with two revolving disks, one covered with photographic emulsion and the other perforated by radial slits. When triggered, the disks revolved in much the same way that a revolver rotates bullets through its chamber. Marey's "rifle" could be trained at a moving object like a gun tracking a bird, and could capture a sequence of images precisely documenting the passage of the object in flight.

After purchasing a long piece of Eastman's transparent film as soon as it came on the market, Marey used it to photograph some pictures of a horse in motion, and even managed to project a few dim, flickering images onto a screen. But he couldn't figure out a way to transport the film through the camera, making this first film screening a bit of a bust.

When Edison returned from Paris, Dickson and his staff greeted their boss with a surprise: a motion-picture projection featuring Dickson himself, stepping out from the screen, raising his hat, and bidding his boss, "Good morning, Mr. Edison, glad to see you back. I hope you had a good time and like the show."

Pleased or not by this demonstration, Edison ordered the primitive technology used to project those images immediately scrapped, to be replaced by a new kinetoscope blatantly based on Marey's "rifle." Dickson promptly wrote Eastman requesting a "cut roll of your Kodac [sic] transparent film 3/4" wide," cut into fifty-foot strips. On November 20, having received no response, he again wrote impatiently to Eastman:

> I wrote some days ago asking you to please send me six 3/4" wide strips 54-feet long or more. . . . I shall require a considerable number of these rolls should my experiments turn out O.K. I am especially anxious to secure from you a non-frilling film which I don't doubt but that your present film proves to be. . . .

That month, Dickson or Edison—most likely Dickson—made the one contribution to the technical development of the motion picture indisputably ascribable to the Edison laboratory. Their solution to the problem of film transport, which had so bedeviled poor Marey, was elegantly simple: They punched holes in the film. Edison's perforations were located at precise intervals to permit entry of double-toothed sprockets, which conveyed the film smoothly through the gate holding the light source.

Though Edison partisans would later stoutly maintain that the width of his film gate was purely arbitrary—that Edison had held his thumb and forefinger apart and said "about yay wide" in response to a question regarding its intended width—thirty-five millimeters happens to be precisely half the width of Eastman's seventy-millimeter film. Dickson and Edison took standard fifty-foot rolls of Eastman's transparent film, slit them down the middle, and cemented the two pieces end to end to form hundred-foot-long rolls.

On May 20, 1891, Edison's wife invited 147 delegates of the National Federation of Women's Clubs to a leisurely lunch at their country estate, Glenmont, which lay just up the hill from Thomas's sprawling laboratory. After dessert, tea, and coffee, the ladies were escorted on a tour of the lab, which included a special surprise treat: a sneak preview—one at a time, please—of Mr. Edison's brand-new kinetoscope.

"Through an aperture in a pine box standing on the floor, the observers could make out the picture of a man," reported the *New York Herald*. "It bowed and smiled and waved its hands and took off its hat with the most perfect naturalness and grace." The man in the box, bowing and scraping appropriately, was William Laurie Kennedy Dickson, wearing a straw hat, dark vest, and white shirt. Dickson's career as a movie star was fated to be as fleeting as Occident's. Five months later, following in the resigned footsteps of Henry Reichenbach, the true "inventor" of the kinetoscope dutifully signed the copyright of his invention—"an improvement in the art of photography"—over to the Old Man.

As Edison was gearing up to send his kinetoscope into the market, he received a curious note from Eadweard Muybridge, wondering how the motion-picture work was proceeding. Edison cagily replied, "I have constructed a little instrument which I call a Kinetograph. But I am very doubtful if there is any commercial future in it and fear that [it] will not earn back its cost. These Zoetropic devices are of too sentimental a character to get the public to invest in."

Despite this obvious attempt to throw Muybridge off track, Edison moved aggressively forward on commercializing his peep show. On April 6, 1894, he secretly shipped twenty-five kinetoscopes from his factory in Orange to Chicago, Atlantic City, and New York City. Ten went to Herald Square, where the finishing touches were being put on a former shoe shop soon to gain fame as the world's first kinetoscope parlor.

For twenty-five cents, viewers bought themselves a chance to peek into five pine boxes arranged in a row, in which fifty-foot film strips of perforated celluloid manufactured by Eastman's rival Blair Camera Company (one of the few companies willing to take on the Eastman juggernaut). The film advanced from reel to reel, just as it had in Marey's photographic rifle, but with the precise alignment afforded by Edison's patented hole-and-sprocket transport arrangement. When a patron dropped his five nickels into the slot, a battery-powered motor moved a strip of celluloid film past a magnifying lens and an electric lamp at the rate of forty-six frames per second.

Not being a professional theatrical man, Edison had handed over the exhibition operation to two Ohio-based vaudeville impresarios, Norman Raff and Frank Gammon, who in return for the right to market the kinetoscope agreed to pay Edison a handsome royalty on just about everything. Edison personally produced not only the exhibition equipment but also the motion pictures. But in keeping with the normal routine at his industrial atelier, the real work of directing the motion pictures was accomplished by Dickson at a tar-paper-sided studio in Orange nicknamed Black Maria, due to its resemblance to a police paddy wagon.

In their first year of operation, Edison's kinetoscope parlors were a smashing success, hauling in millions in nickels and quarters, with more opening every day across the country, amid plans to expand internationally. But Edison's high-handed ways, featuring a $200 price tag on each kinetoscope in addition to hefty fees for every motion picture rented, were the seeds of his undoing.

When Antoine Lumière—a former poster painter whose dry-plate establishment in Lyons, France, was second in size only to Eastman's in Rochester—looked over Edison's fee chart to open a kinetoscope parlor in Lyons, his face darkened into a Gallic scowl.

"Mon Dieu!" he exclaimed to his mechanically inclined son Louis. "Can't you come up with something better?"

Louis certainly could. His brother Auguste later claimed that Louis invented the cinema during one sleepless night, when he was kept up tossing and turning by distressing nightmares and a migraine headache. Louis later discounted the originality of his invention. "What did I do? It was in the air! Other machines preceded mine. Of course," he added dryly, "*they* didn't work."

Lumière retained Edison's perforations but reduced Edison's film frequency from forty-six to sixteen frames per second. This was still fast enough to achieve "persistency of vision" but slow enough to take advantage of forty-six times more light than Edison's kinetoscope. Lumière had a radically different notion from Edison about what the people wanted to see—influenced by Marey, he believed that they yearned to see images projected on a great screen, like the dioramic scenes his father had painted as a young man. They were desperate for a new theatrical experience, not a one-on-one peep show.

Lumière popularized and democratized the motion picture by drastically reducing the cost and the weight of equipment required. Instead of having to outfit a large room with dozens of pine-box machines, Lumière could delight all comers with one small machine: a projector. With an intuitive understanding of the psychology of crowds and the delight people have always taken in all watching the same thing at the same time, the Lumière family proceeded in secrecy to hire out the basement of the Grand Café on Boulevard des Capucines, the exotically named Salon Indien.

On opening night, December 28, 1895, the Lumières' *cinématographe* lured thirty-eight curious souls down the awkward spiral staircase to the basement salon. Each paid a franc for the chance to witness a sampling of astonishingly simple fifty-foot feature films, drawn from the daily life of the Lumière family.

"Baby's First Meal," "Lunch Hour at the Lumière Factory," and "The Sprinkler Sprinkled"—a rollicking comedy in which a gardener is hilariously squirted by water from his own hose—were all enormous comic successes. What occurred that opening night was profoundly different from anything that had ever taken place at Edison's dour kinetoscope parlors. When something funny took place on the screen—and what could be funnier than a baby's first meal?—they *all* laughed. Together.

"Arrival of a Train," which documented the entry of a train into Paris's La Ciotat station, turned out to be something entirely different: the world's first horror film and the season's sleeper. As the great steam-spewing locomotive seemed to burst out of the screen and leap out over the rows of seats into the laps of the audience staring wide-eyed below, nobody laughed. They screamed, leapt out of their seats, charged down the aisles, and shouted at the tops of their lungs, nearly causing a riot as the fear spread like a contagion, just as the laughter had spread.

As a rousing conclusion to the night's offerings, in a solemn homage to the father of the French cinema, and a gentle slap at Edison, the Lumières showed the first newsreel, featuring a group of delegates to a Lyons photographic contest stepping down a gangplank onto the deck of a docked ship. The remarkable thing about Louis Lumière's camera was that it was portable, handcranked, and did not—like the kinetoscope camera— need to be powered by electricity. The first camera capable of filming news events as they broke closed in on the face and figure of Étienne-Jules Marey, the "true"—that is, French—inventor of the motion picture.

The Lumières celebrated the new year (1896) by taking their show on the road—to London, Vienna, Geneva, Madrid, and Belgrade. Royalties, celebrities and nobility flocked to the Lumières' *cinématographe*, thrilled by the novelty of its wide-screen projection, thrilled to be seen in each other's company, thrilled at the newsreels pioneered by the Lumières, thrilled at the gala social event of it all—this was true theater, not some dingy parlor that looked like a betting shop.

With every glowing report of the Lumières' stunning success on the Continent, Edison franchisees Raff and Gammon's mood darkened. They dreaded the day when the Lumières opened in America. As Gammon advised an anxious colleague, "our candid opinion is that the Kinetoscope business will be a dead duck by the end of the season."

Raff and Gammon knew they only had a couple of months, at best, to come up with a movie projector capable of heading the galloping Lumières off at the pass. But the bullheaded Edison refused to believe that the Lumières' projection format was anything more than a cheap novelty. An isolated man, never fully capable of grasping how or why people communicated—he had originally regarded the phonograph as no more than a Dictaphone—Edison simply could not sanction the idea that people would want to watch motion pictures all packed in together in rows of seats in a theatrical hall. That smacked of vaudeville, and burlesque, while in his starch-collared high mind, a good movie was meant to be watched, respectably standing up, appreciated in the privacy of one's mind.

Meanwhile, Raff and Gammon were desperately scouring the

trade shows for a contender to put up against the heavyweight champ: the *cinématographe*. While attending an industrial exposition in Atlanta, they witnessed a demonstration of America's only answer to the Lumières, the phantoscope. Invented by machinist Francis Jenkins and exploited by Washington real estate operator Thomas Armat, compared to Edison's dainty little pine box the phantoscope was a three-hundred-pound gorilla.

But it was Raff and Gammon's last chance. Over Edison's strenuous objections, the Edison Company bought the rights to the phantoscope, renamed it the vitascope, and desperately promoted it as "The Wizard Edison's Latest Invention." Only Raff and Gammon and a few top men in Edison's organization knew the humiliating truth: that the vitascope was the curmudgeonly Old Man's technological "cry uncle."

Fortunately for the Edison interests, the vitascope enjoyed a triumphant debut in April of 1896 at Koster & Bial's Music Hall in Herald Square, on the site of the present-day Macy's department store. Movie audiences, long since tired of Edison's peep shows but not yet exposed to the wonders of the *cinématographe*, thrilled to "Edison's Triumph," as the *New York Times* described it.

Little did they knew that this was only a pale imitation of the Lumières' extravaganza, soon to be coming to a theater near them. Entirely contrary to Edison's firm conviction that audiences would quickly "grow bored" watching movies on a screen, huge crowds reacted ecstatically to the simplest shots of waves crashing on Dover Beach.

"Spectators shouted and squirmed in their seats," the *Times* reported, clearly convinced that they were about to be drenched by the waves.

The vitascope and the *cinématographe* both depended upon celluloid film, punched with holes and fitted with sprockets, to reproduce visual reality in a form so realistic that people physically reacted as if what was happening on the screen was actually happening to them. The willing suspension of disbelief underpinning all theater was not, in this case, willing at all but utterly involuntary. Here, at last, was plastic reality, a hyperkinetic reality available to all, infinitely replicable.

The vitascope put Edison back in the ball game. But only for a few months, until August. In the midst of a record-breaking heat wave, the Lumières opened at Keith's Theater in Union Square. The heat hardly kept the audiences away, as fickle film fans, tired of watching waves roll in on Dover Beach, defected in droves from Herald downtown to Union Square.

This was no director's idea of entertainment but the hard news, presented in blazing if jerky black-and-white motion. The Lumières took the crowd to see—as if they were there!—the coronation of Czar Nicholas II in Moscow, Queen Victoria's Diamond Jubilee Procession in London, and in the latest demonstration of their technical prowess, "Venice As Seen from a Moving Gondola."

For Edison, the Lumières' resounding victory was bad enough. But the worst was soon to come. A few months after the *cinématographe* demolished the vitascope in open competition, a second American projector, the biograph, took its first bow at Hammerstein's Opera House in Times Square.

The biograph projector was no more than a cheap knockoff of the Lumières' *cinématographe.* But it wasn't that which bothered Edison so much as the fact that it was the brainchild of his old minion, William Laurie Kennedy Dickson. Disgusted with Edison's refusal to let Dickson develop a true projector to compete with the looming Lumière threat, humiliated by the purchase of the inferior phantoscope, Dickson had decamped to join Biograph, the one company willing to take on the Edison interests on his home turf.

Biograph's projector opened shortly after the New Year, 1897, exhibiting mainly newsreel footage filmed by Dickson, like the Lumières', but more action oriented. Dickson's stunningly original combat footage of the Boer War would prove to be the most extraordinary achievement of an otherwise checkered post-Edison career. The biograph remained at the Union Square theater until 1905, making its eight-and-a-half-year tenure the longest run in American theater history.

But Edison was legally prepared to take on all comers, armed with a basic motion-picture patent obtained in August 1897, the year of biograph's terrible triumph. This patent permitted him to sue all rivals—including the traitor Dickson—for patent infringement. Furious at the commercial success of biograph and *ciné-*

matographe, Edison was determined to exact justice in the courts for losing the monopoly he had failed to maintain in the marketplace.

Just as Edison was preparing to go to trial, a tragic fire in Paris produced the first real-life horror film. In an improvised canvas-and-wood tent-auditorium thrown up to shelter the Bazar de la Charité (one of the city of light's leading charitable events) an ether lamp providing illumination to a Lumière projector exploded, causing the exceedingly inflammable celluloid film to burst into flames. Over a hundred of Paris's aristocrats died in the conflagration. Those not burned in the fire itself were trampled to death by the fleeing crowd.

The Bazar de la Charité fire was a devastating setback for the French cinema, turning a swank evening's activity, eminently suited for aristocratic diversion, into a popular pastime scrupulously shunned by the well-born. For decades in France, "cinema" would be stigmatized as a dangerous, low hobby, patronized by members of the lower classes foolish enough to risk life and limb for an evening's cheap entertainment.

Which only proved that there was something inherently democratic about celluloid. Originally developed as a means of bringing an elite activity, billiards, to the masses, celluloid had democratized the still image by removing it from the refined world of painting—the traditional province of the aristocracy—and placing it in the hands of everyone. Once Edison/Dickson punched holes in celluloid film, the moving picture became—just like the celluloid collar and the celluloid tortoiseshell cigarette case—a medium for the mass, not just the upper class.

Back in America, Edison was preparing to rein in a rebellious film industry by hauling all the recalcitrant producers into court for patent infringement. After being handed a resounding victory on the first round, in March 1902 the U.S. Court of Appeals pronounced its final verdict in favor of the public at large.

> Mr. Edison . . . was not a pioneer in the large sense of the term, or in the more limited sense in which he would have been if he had also invented the film. He was not the inventor of the film.

Undoubtedly Mr. Edison, by utilizing this film and perfecting the first apparatus for using it, met all conditions necessary
for commercial success. This, however, did not entitle him
under the patent laws to a monopoly of all camera apparatus
capable of utilizing the film. . . .

Following his disastrous defeat at the hands of the second highest court in the land, Edison threatened to pursue his suit all the
way to the Supreme Court. But even Edison's worst enemies were
beginning to grow weary of the incessant litigation and to fear that
the prevailing adversarial atmosphere was hampering the burgeon-
ing industry's growth.

Undaunted, Edison continued to insist on controlling the film
industry the only way he knew how: by exerting legal control over
the film stock itself. He proposed that the producers pay him a fee of
one-half cent for every foot of raw film stock consumed. Film stock
would be handed out only to Edison cartel members who agreed to
distribute films only on stock manufactured exclusively by Eastman.

This cozy arrangement ultimately broke down. The strong-arm
tactics employed by Edison's Motion Picture Patent Company to
keep producers in line forced those unwilling or unable to pay up to
decamp from the East Coast to such exotic locations as Hollywood,
Florida—and Hollywood, California.

For decades Edison never forgave Dickson his treachery. After
receiving a letter from Dickson (dated January 19, 1926) asking
him to vouch for his loyalty as an employee, Edison scrawled in the
margin a curt note to his secretary:

"He was disloyal. He Double X'd me. I think we'd better not
answer. File."

The Bakelite Brigade

Leo Baekeland
(Courtesy of Union Carbide Corporation)

The solid citizens of Yonkers were never quite sure what to make of "Doc" Baekeland, whose sole recreation seemed to be driving through the quiet streets of their sleepy Hudson River town in his motorcar, a contraption still widely denigrated as noisy, dirty, and dangerous in the early years of the American century.

Most residents of the still rural New York suburb were familiar with the story of how their brilliant young neighbor, a Belgian immigrant, had set himself up at Snug Rock—a handsome estate set high above the Hudson River, commanding a sweeping view of the Palisades across the way—at the precocious age of thirty-five. Ten years after his arrival in America, a photographic paper he developed caught the eye of George Eastman, who invited Baekeland to Rochester to discuss a possible sale. On the train up from New York, he decided to ask for $50,000. In a pinch, he'd settle for half.

No matter what inducements or blandishments Eastman might dangle before him, Baekeland was resolved to stand firm at twenty-five. Once seated in Eastman's office, however, Baekeland found himself with precious little room to maneuver. After bestowing high praise on his entrepreneurship—the young man had not only invented the new paper but had overcome considerable resistance on the part of photographers in marketing it—Eastman made a preemptive offer to relieve him of the burden of running his six-year-old Nepera Chemical Company. Assuming, of course, they could come to an agreement on price.

Baekeland scarcely had a chance to open his mouth before Eastman casually tossed out a number. Baekeland wasn't sure that he had heard right. Could Eastman have possibly asked, "How does three-quarters of a million strike you?" The astonished inventor scarcely had time to pick his jaw off the floor before the older man leapt up from his seat and began eagerly pumping his hand. Baekeland rode home to Yonkers wealthy beyond the wildest dreams of a struggling immigrant who had named his first son George Washington out of gratitude to a land that delivered on its promises. On the other hand, Baekeland was now out of a job.

> At thirty-five I found myself in comfortable financial circumstances, a free man, ready to devote myself to my favorite studies. Then truly began the very happiest period of my life. I improvised one of the buildings at my residence in Yonkers into a modest but conveniently equipped laboratory. Henceforth, I was able to work at various problems of my own free choice. In this way, I enjoyed for several years that great blessing, the luxury of not being interrupted in one's favorite work.

Celine Baekeland (whom everyone knew as Bonbon) played a key role in providing her husband with the luxury of not being interrupted in his favorite work by loudly stage-whispering to their daughter Nina, whenever her boisterous playing threatened to disturb his sacred peace, "Shh, dear! The Doctor's working!"

As to what precisely the Doctor was working *on*, only Leo and his assistant knew for sure. Celine occupied herself with tending to her prize roses, painting in oils, and conducting intimate musical soirees in their elaborate drawing room. "Doc" seemed well suited to the role of wealthy self-made businessman, on a par with the other privileged gentlemen from their leafy suburb, Harmony Park, who dutifully donned dark suits, high collars, and ties every morning to board the Hudson River Line at Yonkers station, returning every evening on the dot of 5:40, like automatons in a time-motion film sequence by E. J. Marey. But Baekeland's office lay close at hand, directly across his back lawn in a converted stable, a place that his neighbors couldn't help noticing exuded peculiar smells on warm summer nights, depending upon which way the wind blew off the river. At times, the place smelled like a mortuary; at other times, like a hospital.

Most of those affluent neighbors would never have guessed from his well-tailored suits and solemn, old-school demeanor that Leo had grown up a poor boy in the old country, the son of a cobbler and a maid, both illiterate, from Ghent in Flemish Belgium. Fortunately for Leo, during years of service in the homes of the well-to-do his mother had acquired an appreciation for the finer things in life.

Convinced that a fine education was a passport to a better future, Rosalie Baekeland (whom everyone knew as "Merchie') saw to it that her son was enrolled in Ghent's top grammar school. Leo helped pay his school bills by apprenticing himself in his father's shoe shop. But after a few years of honing his leathercraft skills in preparation for a future he hoped would never take place, the municipal authorities granted him a City Scholarship, which permitted him to enter the Royal Athenaeum, breeding ground of Ghent's elite. The act of relieving his parents of the burden of supporting him was deeply gratifying to Leo, who had been reading a life of Benjamin Franklin. Young Ben had also freed himself from parental dependence at an early age, setting a stirring example Leo felt compelled to follow.

After a day's worth of courses at the Athenaeum, Leo spent his evenings studying and attending lectures in chemistry at the local technical school. Still, he found time to indulge in photography, an extracurricular passion strongly encouraged by Dr. Desire Van Monkhoven, a local chemist who had written several books on photography and founded a small factory in Ghent to manufacture photographic dry plates—the first in Europe.

Behind a solemn demeanor Leo concealed a mischievous streak, which slyly expressed itself in a compulsion to play practical jokes. One day, bored by class, he decided to demonstrate the indelible properties of silver nitrate on the face of one of his schoolmates. After catching a glimpse of his sun-printed facial photograph in a mirror, the boy panicked and ran to his family doctor, who prescribed hydrochloric acid as a stain remover.

Alarming results—burns but no scars—prompted a school disciplinary committee to look into the incident. Rather than speak out in his own defense, Leo responded to his examiner's questions by calmly swabbing some silver nitrate on his own face, and just as calmly wiping it off with a harmless chemical he produced from a vial concealed in his coat pocket. Baekeland's eloquent pantomime prompted the board to let him off with a slap on the wrist. The poor doctor was not so lucky—his professional reputation was permanently tarnished.

After graduating from the Athenaeum at seventeen, Leo enrolled at the University of Ghent, where he was celebrated as the youngest student on campus. He soon attracted the attention of senior chemistry professor Theodore Swarts, the great Friedrich August Kekulé's chief assistant when he had taught at Ghent two decades before. Kekulé had announced his classic theory of the structure of benzene at Ghent before moving on to a more prestigious post at Bonn.

Baekeland sailed through university in two years, earning his B.S. degree with highest honors in 1882 and a doctorate two years after that. After a brief stint teaching at Bruges, about twenty miles from Ghent (where Professor Swarts also headed the chemistry department), he returned to his alma mater as an associate professor, to pursue a blooming love affair with Celine Swarts, the beautiful daughter of his senior professor. When he first met Celine in her father's laboratory, he had been so violently smitten that he dropped a pair of beakers at her feet.

"My most important discovery at the University," Baekeland gallantly recalled, "was that my senior professor of chemistry had a very attractive daughter. Hence, the usual succession of events . . . "

But the course of true love was not destined to run smoothly. A confirmed shutterbug, Leo was convinced that he could produce better dry plates than his mentor Monkhoven, and launched a private business to do so. Aware that most photographers would rather not bother with smelly, tricky chemical developers, Baekeland developed a dry plate coated with a water-soluble emulsion, which only had to be dipped in water to activate the developer beneath.

With a friend from the university faculty, Baekeland founded a partnership, Baekeland et Cie., and leased small quarters in a warehouse to produce water-activated dry plates in bulk. But he found himself spending all his spare time struggling to keep it afloat, and in constant conflict with Professor Swarts over his involvement in business affairs.

Celine's father did not understand what could possess a brilliant young man like Baekeland, with a shining academic career before him, to trifle in commerce. Though Celine and Leo were married in August 1889, the wedding took place against the professor's wishes in an atmosphere of quiet acrimony. The week after their wedding, the couple took off for Oxford and Edinburgh on a traveling fellowship Leo had been awarded two years before. From London they set sail for New York, where the young inventor hoped to forge some profitable ties with leading American photographic suppliers.

As the esteemed young inventor of the first water-developable dry plate, it didn't take long for Leo to be introduced to Richard Anthony, president of A. & H. T. Anthony & Co., an old-line photographic supply house. After quizzing Baekeland on some questions involving "chemical problems connected with the manufacture of bromide paper and photographic films," Anthony came away so impressed that he offered Leo a job as a staff chemist, effective immediately.

At twenty-seven, having only just arrived in New York, Baekeland had to carefully calculate the trade-offs between the safety, security, and prestige of an academic career in Europe and an opportunity to strike out on his own in the New World. Before the turn of the century, industrial chemists were virtually unknown in the United States. Though Du Pont had founded a research laboratory at Carneys Point, New Jersey, in 1891, its experimental work was strictly limited to the field of smokeless gunpowder. General Electric's research laboratory at Schenectady, generally considered the first industrial research laboratory in the United States, was founded in 1900—a year after Baekeland's New World debut.

Leo didn't take long to weigh Anthony's offer, but promptly cabled his resignation to the Belgian minister of education, who, after accepting his resignation with regret, conferred upon him the honorary title of associate professor at Ghent in the hope that he might one day return home to roost.

Turning his back on his budding European academic career, he fancied himself a self-taught maverick in the rough-and-ready manner of Edison, Eastman, and Hyatt, as opposed to a highly trained academic sophisticate with a top-notch Continental education behind him. Baekeland's compelling need for independence prompted him to leave Anthony & Co. after only two years and take the risky step of hanging out a shingle as an independent chemical consultant at a time when freelance chemists were few and far between.

The turn of the century was a feverish time to embark on a career at the frontier of science and technology. The compelling myth of the wild, rugged individualist (so energetically embodied by President Teddy Roosevelt) had a curious scientific counterpart in the lab. Just as earthbound explorers were compelled to conquer new territories in Africa and Asia by dint of wile and strength of character, the lone genius of science was expected to heroically thrust back the frontiers of human knowledge. As Baekeland later insisted:

> My real education began only once I left the University, as soon as I became confronted with big problems and responsibilities of practical life. . . . What little education I received was mainly in the United States. I hope to remain a postgraduate student in that greater school of practical life, which has no fixed curriculum and where no academic degrees are conferred, but where wrong petty theories are best cured by hard knocks.

He spent the next several years "trying to work out several half-baked inventions, the development of each of which would have required a small fortune." Lacking that, he floundered professionally until, seriously ill, he was forced to take to his bed for several months with nothing to do but ponder his plight.

> While hovering between life and death, with all cash gone, suffering from the uncomfortable sentiment of rapidly increasing debts, I had abundant time for sober reflection. It then dawned upon me that instead of keeping too many irons in the fire, I should concentrate my attentions upon one single thing, which would give me the best chance for the quickest possible results.

His health restored, Baekeland emerged from his sickbed rejuvenated by the realization that the one thing that *might* hold out the promise of tangible results was an idea he had entertained for ten years: a more light-sensitive photographic paper. For the first half-century of photography, printing had been accomplished by sunlight. But Baekeland was convinced that he could prepare a silver chloride emulsion in a "colloided" (jellied) state that would eliminate the need to wash and "ripen" (heat up) the paper, the only means then known to boost light sensitivity.

Baekeland's jellied-emulsion-coated paper turned out to be so much more light sensitive than other papers that it could be developed by artificial light. In 1892, that meant gaslight. But by threatening to put an end to the cherished tradition of "sun printing," Baekeland was committing "photographic heresy."

Compounded by a devastating financial crisis that swept through the country in the spring of 1893, unanticipated resistance on the part of professional photographers nearly did Baekeland's company in before it was up and running. Teetering on the verge of bankruptcy, Baekeland launched a last-ditch effort to target amateur photographers, who might at least be "willing to take the trouble to read and follow our printed directions."

Fortunately, open-minded amateurs couldn't get enough of Velox. With the new paper, they could develop photographs quickly and with relative ease at home, or hand them over to a new crop of low-cost neighborhood labs that had sprung up to take advantage of the new high-speed photo-processing. Inside two years, the rampant enthusiasm of amateurs made Velox a household name in the industry. When George Eastman made Baekeland an offer he couldn't refuse, his opening bid—roughly $25 million in today's dollars—could be justified only because Baekeland had built his Velox-brand paper into a thriving franchise.

The million-dollar sale to Eastman Kodak provided Baekeland with better advertising than any money could buy. At precisely the point when he no longer needed the work, offers flooded in to retain him as a consultant, at any price he cared to name. After returning in triumph from Niagara Falls, where he helped Elon Hooker, founder of Hooker Chemical (future polluter of Love Canal) build the largest electrolytic-cell-power facility in the world, Baekeland retired to his Yonkers abode, eager to once again settle down to his "favorite work"—finding intractable chemical problems in need of innovative solutions.

For months, he toyed around with a method of making soybeans tastier and with a way of disinfecting wood with bisulfate liquor. But isolating that *one single thing* that would give him "the best possible chance for the quickest possible results" eluded him for over a year. Only after anxiously scanning the scientific literature and jotting down innumerable trenchant observations in his notebook did he finally arrive at the conclusion that what the world then needed most was a viable synthetic shellac.

Just as a precipitous decline in the elephant population had sent billiard-ball ivory out of reach of the average citizen, the rapid expansion of the electrical industry had caused the price of shellac—derived from the resinous secretions of the female lac beetle—to soar above a dollar a pound. As long as shellac's only widespread application was as a lacquer and preservative for wood products, traditional Third World production methods were efficient enough to meet routine demand. But by the turn of the century, the United States alone was importing upwards of a million pounds of shellac a year.

When a female *Laccifer lacca*, a species of beetle native to Southeast Asia, attaches herself to the bark of a tree, she sucks sap from the wood and excretes a resin that gradually accumulates until it encases its host in a transparent amber-colored shell. After reproducing itself a couple of times, the beetle dies, having incarcerated itself in its own excrement.

After gathering the dead shells from the trees, Indian peasant gatherers had for centuries spent untold hours melting the shells over wood fires in iron pots, painstakingly filtering the amber-colored hot liquid to remove the leaves, bark, and insect bodies preserved inside. Since it took fifteen thousand lac beetles six months to produce enough resin to yield a single pound of shellac, the cost of the one truly viable electrical insulator available had suddenly soared through the roof.

In 1902, the thirty-eight-year-old millionaire entrepreneur Leo Baekeland embarked on a scientific adventure destined to make his the first household name of the Synthetic Century. The prime subject of his five-year exploration did not, at first glance, merit intensive attention. Thirty years earlier, the eminent German chemist Adolf von Baeyer, famous for synthesizing indigo, had set off a violent reaction between phenol—a coal-tar distillate used as a turpentine substitute—and formaldehyde, a wood-alcohol distillate used by morticians as embalming fluid. But after obtaining a tarlike solid that appeared utterly impervious to chemical analysis, Baeyer derisively dismissed it as nothing more than a mere *Schmiere*.

Phenol-formaldehyde potions might easily have remained consigned to scientific obscurity if a cat residing in the laboratory of a Bavarian chemist, Adolf Spitteler, had failed to knock over a bottle of liquid formaldehyde, which spilled into her saucer of milk. To Spitteler's astonishment, the milk quickly curdled into a hard, hornlike compound resembling celluloid. The academic chemist would never have dared to brave the torrents of the open marketplace if he had not been approached later that year (1897) by Ernst Krische, a Hanover-based manufacturer of student blank books seeking to fill a large order placed by a German school system for a washable white writing board.

Krische had tried coating cardboard with casein—the primary protein in milk. It had worked like a charm, except that wiping the board down with a damp cloth caused the surface to soften. By adding formaldehyde to the recipe, Spitteler and Krische succeeded in filling the order for a casein-plastic white writing board.

They also succeeded in producing the first commercially successful plastic since celluloid. Marketed in Germany under the trade name Galalith (from the Greek *gala* for "milk" and *lithos* for "stone": milkstone) and in England as Erinoid (a play on the Gaelic place-name of Ireland, since the casein curds hailed from Limerick) casein plastic never posed any great threat to celluloid. But it held its own for years against flashier synthetics as a raw material for buttons because it so easily took to a wide range of colors and could be produced cheaply from humble dairy by-products.

If Krische and Spitteler had not been successful with their formaldehyde-based plastic, Werner Kleeberg, another German chemist obsessed by coal tar, would probably not have bothered to take a second look at Baeyer's mere *Schmiere*. Kleeberg used hydrochloric acid as a catalyst, thereby obtaining a black, sticky, smelly, amorphous substance similar to Baeyer's, but with a key difference: It was insoluble—not dissolvable in the usual solvents—and infusible—impossible to melt.

Kleeberg lacked the foresight to see these traits as exploitable commercial properties. Baekeland later smugly observed, "After failing to crystallize this mess, or purify it . . . or in fact do anything with it, he described it in a few lines, and dismissed the subject. . . ." But in 1902, the same year Baekeland launched his own investigation conceived along similar lines, an Austrian chemist, Adolf Luft, decided that Kleeberg might have been onto something. Luft mixed phenol and formaldehyde in a determined attempt to develop a celluloid-like hard plastic. But the transparent, amber-colored solid he obtained was, in Baekeland's words, "relatively brittle, much less tough, and far less flexible than celluloid."

British electrical engineer James Swinburne took one look at Luft's stuff and declared, "It looks like frozen beer!" Well aware that its high heat resistance made it a promising candidate for a synthetic high-voltage electrical insulator, in 1904 Swinburne formed The Fireproof Celluloid Syndicate Ltd. to market Luft's frozen beer to industry. But since it remained too brittle and unstable to be used commercially, Swinburne spent the next three years searching for a solvent to give it backbone.

Having satisfied himself that caustic soda was the solvent in question, Swinburne set off in a state of carefree anticipation for the British patent office to file his claim. But only hours after his arrival, he was told that he had been anticipated in his research by an American rival named Leo Baekeland. By *one day*.

In 1902, Baekeland had begun modifying Baeyer's and Kleeberg's phenol-formaldehyde reactions by trying to control the violent buildup of gases. Adding acids to these volatile mixtures produced solids that were both soluble and fusible—dissolvable in solvents, meltable by heat—and often spongelike and syrupy. Adding bases like caustic soda or ammonia, however, caused the reaction to drastically accelerate while lowering the gaseous activity, producing solids that, like Kleeberg's and Luft's, were resistant to solvents and heat.

After spending five years trying to control the reaction by slowing it down, in early 1907 Baekeland had an epiphany: Rather than trying to lower the violent buildup of gases by slowing the reaction down, why not speed it up? His predecessors had sought to keep temperatures low as a means of preventing the buildup of turbulent gases. But Baekeland turned the heat up—to well over 200 degrees centigrade—and instead controlled the volatile reaction by increasing the pressure inside the vessel.

He developed an apparatus called a Bakelizer—resembling a copper-pot still, outfitted with a steam jacket and a sealed top — with which he was able to precisely vary heat and pressure, exerting a considerable degree of control over the resulting reaction. After a few trial runs of the Bakelizer, Baekeland moved quickly and confidently toward his ultimate goal.

On June 20, 1907, he noted with mounting excitement the presence of:

> A solidified matter, yellowish and hard ... This looks promising and it will be worthwhile to determine how far this mass ... is able to make molded materials, either alone or in conjunction with other solid materials, as for instance asbestos, casein, zinc oxide, starch, different inorganic powders and lamp black and thus make *a substitute for celluloid and for hard rubber.*

His sole diary entry for June 18–21:

> I consider these days very successful work which has put me on the track of several new and interesting products which may have a wide application as plastics and varnishes.

In one intense week of practically around-the-clock work, Baekeland completed his five-year quest in a sober, beeline dash for the finish line. His final product was a liquid (Bakelite resin) that rapidly hardened into a transparent, amber-colored solid, forming an exact replica of the vessel used to contain it. And its surface was engraved, as if by a master craftsman, with the precise pattern of the vessel's every seam and flaw.

On July 14, he took out his famous "heat and pressure" patent:

> Be it known that I, Leo H. Baekeland, a citizen of the U.S. residing at Snug Rock, Harmony Park, Yonkers, have invented certain new and useful Improvements in Methods of Making Insoluble Condensation Products of Phenols and Formaldehyde. . . .

Unlike shellac, Bakelite wouldn't burn, boil, melt, or dissolve in any acid or solvent commonly available. And though it could be produced in liquid form, once "frozen" by heat and pressure it retained its solid form even if reheated. As opposed to a "thermoplastic" compound like celluloid, which can be reheated and remolded in an infinite cycle, Bakelite was the first "thermoset" plastic. Which meant that once it was firmly set, it was set for life.

More significant scientifically, Bakelite was the first true synthetic material. As opposed to celluloid (a natural polymer composed of cellulose molecules modified by nitric acid and camphor), Bakelite was fashioned from man-made molecules that have no precise duplicate in nature.

On February 5, 1909, at the Chemist's Club in New York, Baekeland, not by nature a modest man, introduced Bakelite to the world. As the *New York Sun* confirmed in a headline, "Baekeland Claims Much for New Invention." Not only could he "take a piece of wood, dip it into the liquid resin, place it in the Bakelizer for an hour or so, and provide it with a hard, brilliant coat of Bakelite better than the best and most expensive Japanese lacquer," but—by plunging a piece of cheap, soft porous wood into a Bakelite bath and then hardening it in his Bakelizer—he could magically transform soft wood into hard wood.

Anything nature could do, Leo could do better. Rather than simply improving the surface of the material, Baekeland could chemically improve its inner self.

But Bakelite's reputation was made, as expected, as an electrical insulator vastly superior to any natural material on the market. It was more electrically resistant than porcelain or mica; more chemically stable than rubber; more heat resistant than shellac; less liable to shatter than glass or ceramic; it would neither crack, fade, crease, nor discolor under the influence of sunlight, dampness, or salt air; it was impervious to ozone, contained no sulfur to cause the "greenling" (degradation over time) suffered by hard rubber, and could not be weakened by hydrochloric acid or blemished by alcohol. Contact with oil- or grease-stained fingers would not warp, mar, or disfigure it; it was virtually impervious to natural or human attack.

Due to an elasticity characterized by its inventor as "approaching that of ivory," Bakelite also happened to make perfect billiard balls, fulfilling John Wesley Hyatt's decades-long dream of forging the ideal artificial billiard ball. Bowing to the synthetic inevitable, in 1912 Hyatt personally ordered his Albany Billiard Ball Company to stop using celluloid and substitute Bakelite as its molding material.

Courtesy of Union Carbide Corporation

After suffering the traumas associated with trying to market Velox to an initially indifferent public, Baekeland was justifiably wary of trying to make Bakelite on his own. His first inclination was to grant licenses to selected manufacturers and stay out of the production end altogether. But once it became clear that the Bakelite process was too complicated and cumbersome to be entrusted to ordinary, run-of-the-mill molders, Baekeland found himself churning out hundreds of gallons of Bakelite resin in his garage.

As an early associate recalled:

"Old Faithful" [the first Bakelizer] had an agitator, which required electrical power. But electric light lines had not yet reached Harmony Park. So the Doctor acquired a steam engine from an old White Steamer automobile. The steam engine required steam, so a small vertical coal-fired boiler was erected in one corner of the laboratory, and the steam piped across to the garage. . . .

In March 1909, a flash fire consumed most of the garage, nearly torching the adjacent laboratory. Even more alarmingly, the fire totaled Doc's treasured Peerless automobile in which he had tooled around Europe at the turn of the century—one of the first motorists to do so. When Baekeland learned that the Perth Amboy Chemical Works in Perth Amboy, New Jersey (a major formaldehyde manufacturer), had space to rent in one of its buildings, in 1910 he relocated The General Bakelite Company to safer quarters, much to Celine's relief.

It didn't take long for the not-terribly-catchy slogan "Material of a Thousand Uses," reputedly coined by the inventor himself in an expansive moment, to be confirmed in reality. After calculating in the late teens that forty-three industries directly benefited from Bakelite, by the late 1920s Baekeland proudly reported that he could not name forty-three industries that had no use for it at all.

Jazz Modern

The name [Bakelite] summons up a whole mystique of early plastics for Art Deco collectors, producing images of Cubist-style 1930s' jewelry, mock-tortoiseshell compacts, ersatz ivory combs, butterscotch-toned desk accessories, and what may well be the most glamorous plastic creations of the period—'streamlined style' radios.

RITA REIF, *NEW YORK TIMES*, JUNE 3, 1984

Courtesy of Bell Laboratories

During the summer of 1925, *L'Exposition Internationale des Arts Décoratifs et Industriels Modernes* of Paris arose like some "Cubist dream city, or projection of a possible city on Mars" (as one awestruck American visitor put it) at the blazing heart of the City of Light.

The great exposition's seventy-five-acre site straddled the Seine, sprawling from the Grand Palais all the way to the Place des Invalides. "It looked like the inside of a kaleidoscope," exclaimed another, equally awestruck American visitor, having roamed until his feet nearly gave out through the two-hundred-odd pavilions constructed by manufacturers from seventeen countries, with the notable exceptions of Germany—excluded for political reasons—and America, whose manufacturers had insisted that they "had nothing to offer in the modern mode."

But Europe—particularly France—had modern on the brain. Unlike earlier world's fairs, which had been multipurpose affairs conceived to showcase all the latest products of commercial art and technology, the 1925 "Art Deco" fair in Paris was devoted exclusively to the decorative arts. No central structure, no Eiffel Tower or Crystal Palace, aspired to attain iconic status as an enduring symbol of a new era. Only Le Corbusier's deceptively small, incongruously austere, white steel, glass, and concrete cube commissioned by the avant-garde magazine *L'Esprit Nouveau* (New Spirit) could possibly have qualified. But Le Corbusier's cube was set so far away from the center of the site that it was scarcely noticed by the majority of visitors.

In any event, the crowd vastly preferred the more conspicuous pavilions, the most lavish of which was more a testament to the luxury of a bygone era—although presented in a style that was entirely *à la mode*—than a manifestation of the stripped-down modern life. Executed by the renowned Parisian interior designer Jacques-Émile Ruhlmann in the form of an imaginary French embassy, the exotic hardwood parquet floors, flamboyant white polar-bear-skin and zebra-skin rugs, ornate stained-glass windows, and sleek angular-style furniture modeled after the French Empire style were more of a last languid gasp of the voluptuous style known as Art Nouveau than a shrine to the streamlined Machine Age.

Even Le Corbusier and his fellow Modernists (including the

absentee Germans of Weimar's Bauhaus school) preferred to pay lip service to the *idea* of machinery and mass production while remaining deeply committed in practice to an Arts and Crafts tradition that endorsed radical forms as long as they followed function. The classics of Modernism were nearly all executed in traditional natural materials: fine woods, luxurious leathers. Steel, glass, and concrete were acceptable. But synthetic materials, for the most part, remained in questionable taste.

Only in America could Lewis Mumford praise without irony the mass-produced airplanes, automobiles, locomotives, even the latest kitchen equipment and modern bathroom fixtures, as the essence of a vigorous new American style. While the English cultural arbiter Edith Sitwell summed up the Roaring Twenties in a jazz riff phrase, "the allegro Negro cocktail shaker," in America that cocktail shaker was definitely designed in that swank, smart new material: Bakelite.

After rendering distinguished service in battle during World War I as battleship insulation and artillery-shell casings, a new laminated material called Bakelite Micarta was recommended by Leo Baekeland, in his official capacity as adviser to the U.S. War Board, as a possible substitute for laminated wood in aircraft propellers. A wooden prop, though perfectly adequate for civilian use, had become an Achilles' heel for combat aircraft during aerial dogfights.

Since the advent of the foremounted machine gun, Bakelite engineers cut canvas sheets into wing shapes, cut a square hole to fit the hub, shoved the whole mass in a propeller-shaped mold, stuck the mold into a powerful hydraulic press, and baked the whole sandwich under high heat and pressure. After a few hours, out from the mold popped a perfect propeller, hard as a rock, smooth as glass, with the textured weave of the canvas cloth still visible beneath its icy surface.

After the armistice, Bakelite General heavily promoted the conversion of this tried-and-true combat material for civilian use. The Bakelite propeller was quickly and cleverly adapted to the surging demand for automatic washing machines, which were in high demand since wartime labor shortages had made "good help" so hard to find.

"To sell against the Bakelite impeller is folly," proclaimed the *Electrical Record.*

The Meadows Company is so convinced of the popularity of the molded impeller that in the near future, to sell a washing machine equipped with a metal agitator will be a difficult job indeed.

Meadows promoted its palpably phallic impeller with an ad depicting a smooth, glistening, rock-hard black plastic shaft held smack up to a pert young housewife's blissful face, boldly captioned with the headline "Monday Magic."

"There is no single competitive feature in a washing machine today as intriguing to the feminine public as the slippery Bakelite Impeller," the seductive tag line positively purred. "This amazingly smooth, glossy Bakelite Impeller protects your clothes . . . right down to the filmiest, flimsiest thing!"

In 1920, the first regularly scheduled commercial radio broadcast was transmitted by Westinghouse Electric's pioneer KDKA station in Pittsburgh. After two years of broadcasting, some one hundred thousand radios were in use throughout the United States. But by 1925, the year of the Art Deco fair, annual radio receiver sales had jumped from $2 million to $136 million, with more than 5.5 million receivers in daily use. Those early-model radio sets were vast, cumbersome affairs, solidly constructed of oak and maple veneers to fit in with the heavy Victorian furniture still popular in most middle-class homes.

In 1927, both Macy's and Lord & Taylor's department stores in New York held hugely popular Art Deco exhibitions, while Jacques-Émile Ruhlmann's Art Deco pavilion, on an around-the-world tour, was briefly enshrined in that temple of traditional taste, New York's Metropolitan Museum of Art. Art Deco—known as Moderne in the United States—became all the rage, and Bakelite its defining medium.

During the twenties, Bakelite gained a new distinction as the signature material of a new hard-boiled, hard-drinking "lost generation," the sophisticated, urbane set that devoured the original *Vanity Fair* and giggled at the quirky cartoons in *The New Yorker.* Bakelite conveyed a smooth-yet-solid surface swank sleekly evocative of the Al Capone era, molded into the sweeping sculptural lines of the latest high-speed cruise liner—the stuff of martini dreams. Bakelite was tough and rugged and stark and lean—as stripped down as a Hemingway sentence. When Hemingway admiringly wrote of the stunningly sleek Lady Brett that "she was built with curves like the hull of a racing yacht," she might just as well have been molded from Bakelite Laminated.

By the mid-twenties, those streamlined shapes embodying the shocking look of the new were engineered by a new breed of style-conscious taste setters who called themselves Industrial Designers. Though the Austrian-born Paul Frankl and the French-émigré Raymond Loewy were the only two Europeans in the gang, all the rest—Norman Bel Geddes, Henry Dreyfuss, Harold Van Doren, and Walter Dorwin Teague (collectively known as the Four Horsemen) were refugees from the blatantly commercial worlds of advertising and commercial art.

The Industrial Designers borrowed more from Broadway than from Bauhaus or Beaux-Arts. Their urbane sensibility was shaped more by popular slicks like *Harper's Bazaar, Vogue,* and the new *Esquire* than professional architecture or design publications. They didn't give a hoot if form followed function, as long as style and flair perked up the packaging. Obsessed by surfaces, and by sales, as Raymond Loewy proudly put it, they would take money to redesign anything, "from a lipstick to a locomotive."

By the time *Time*—another recent addition to the mass-media scene—saw fit to put "Doc" Baekeland on its cover for the week ending September 22, 1924, Bakelite had become the latest force of Un-Nature. Beneath the trim-mustached, bespectacled visage of the inventor the caption ran: "It will not burn. It will not melt." *Time* extolled "the Material of a Thousand Uses" as a conquering hero.

It is used in pipe stems, fountain pens, billiard balls, telephone fixtures, castanets, radiator caps. In liquid form, it is a varnish. Jellied, it is a glue. Those familiar with its possibilities claim that in a few years it will be embodied in every mechanical facility of modern civilization.

From the time that a man brushes his teeth in the morning with a Bakelite-handled brush until the moment when he removes his last cigarette from a Bakelite holder, extinguishes it in a Bakelite ashtray, and falls back upon a Bakelite bed, all that he touches, sees, uses will be made of this material of a thousand purposes.

In an ironic play on the name, *Time* wryly predicted:

> Books and papers will be set up in Bakelite. People will read Bakeliterature, Bakelitigate their cases, offer Bakeliturgies for their dead, bring their young into the world in Bakelitters. . . .

What the *Time* cover story neglected to mention was that in 1927, as Art Deco hurtled from coast to coast like the streamlined Twentieth Century Limited, Baekeland's patent protection on the phenolic plastic process was due to expire. The prospect of competition put the heat and pressure on Baekeland to enlist top-notch industrial designers in a no-holds-barred campaign for plastic integrity.

In 1927, the Parker Pen Company of Janesville, Wisconsin, unveiled its new Bakelite barrel. Taking a cue from the stress tests to which the Bakelite propeller had been subjected in wartime, the Parker Pen people created a landmark advertising campaign based on Bakelite's awesome ability to take punishment.

The Duofold Barrel "Stopped Traffic on Fifth Avenue" as a huge (staged) crowd witnessed it being "Hurled 23 Stories to Cement! Picked Up Unbroken!" It was soon to be placed "Under the Wheels of Loaded Buses!" where, like Superman, it was proven impregnable.

Later that year, Bell Telephone offered $1,000 to each of ten designers recommended by a panel of artists to produce designs for a new phone that would combine the speaker and microphone in a single handset. The company's in-house engineers had created an awkwardly functional updated version of the old candlestick model. The one design specification for the new phone was that it be molded from "rugged, durable, phenolic resin." Which meant Bakelite.

Henry Dreyfuss, scion of a family of theatrical designers and one of the leading movers and shapers of industrial design, was one of the candidates approached by Bell on the first round. But Dreyfuss turned Bell down on the grounds that the project demanded a top-to-bottom redesign, a goal that could be accomplished only in close consultation with Bell's own engineers.

After rejecting Dreyfuss's proposal outright, once the contest designs were all in and all deemed unsuitable, a chastened Bell reapproached Dreyfuss, now willing to pay what it took to let him do what he had initially proposed: to fully integrate the phone's new form with its function, from the bottom up. Dreyfuss, for his part, would describe his commission a trifle more archly: "They wanted a little art to wrap the phone in." Dreyfuss's basic black Bakelite rotary-dial receiver (modified 1950) was to define "phone" for decades to come, until the advent of the digital era.

By the late twenties, Walter Dorwin Teague, a former commercial artist and authority on typography, was devoting all his energies to a specialized form of late-Deco industrial design popularly known as "streamlining." Teague's bag was cameras, which he turned into handheld rocket ships under exclusive contract to Kodak. In 1934, Teague's "Baby Brownie," sporting a distinctive vertical-ribbed brown Bakelite housing, sold four million copies in its first year—at the depths of the depression. Teague's "Hawkette" soon followed, rendered in mottled Bakelite. Raymond Loewy weighed in with his own Bakelite Purma Special, whose recessed controls made it a forerunner of late-twentieth-century ergonomic design.

Loewy had been an engineer in Paris before enlisting in the French army in World War I. After the armistice, he set sail for America in pursuit of a wild dream to design modern appliances for General Electric. On board ship, he dashed off a drawing of an elegant young lady promenading on deck, which was purchased for a charity auction by a young Englishman who happened to be British consul in New York.

His new high-society connection introduced Loewy to magazine publisher Condé Nast, who promptly hired him as a fashion illustrator for *Vogue,* which soon lost him to *Harper's Bazaar.* It didn't take Loewy long to become a late-breaking fashion trend. Saks–Fifth Avenue commissioned him to redesign the uniforms worn by their elevator operators, which made his name as a stylish taste setter. But the more successful he became in the frivolous fashion world, the more he yearned to rekindle his old dream of designing appliances for General Electric.

In October 1929, not long before the stock market crash, Loewy was introduced at a London dinner party to a German mimeograph-machine manufacturer, Sigmund Gestetner. The mimeo magnate, suitably impressed by Loewy's eloquent disquisition on the comparative design merits of London versus New York taxicabs, soon paid a call on Loewy at his design studio in New York. He arrived with a clumsy-looking machine in tow, whose four ungainly tubular steel legs stuck out from its base like the splayed shanks of a mule. "As a consumer-conscious designer," Loewy later recalled, "I detected the hazards inherent in the four protruding legs in a busy office."

Gestetner seemed hesitant about giving Loewy the assignment until the flamboyant designer rapidly sketched an image of a shapely stenographer, correspondence in hand, tripping over an obtruding leg, sending papers flying everywhere. Gestetner took one look at the drawing and offered Loewy $2,000 on the spot to redesign the machine from the ground up—in five days.

Loewy spent much of that time covering the mimeograph machine with clay, from which he cast a mold that would encase the machine with a sleek shell like the hull of a yacht. The shell hid the previously exposed levers and gears from view, providing form. But it also had a legitimate function: It kept the inner workings from getting covered in dust, which mixed with printer's ink to form a thick gunk that clogged up the gears.

The Bakelite Company wasted little time trumpeting its latest victory in the design wars currently raging between the Modern and the Traditional. "Raymond Loewy says the 'New Deal' in industrial design is establishing the triumph of beauty through simplicity."

Bakelite even promoted the sublime "union of art and industry" exemplified by designer John Vassos's sleek chromium-banded Bakelite counters at a new Nedick's refreshment stand in New York. "The artist-designer . . . realizes that beauty of form must be expressed in appropriate materials. Like Mr. Vassos, who used Bakelite Laminated . . . "

By 1936, *Fortune* had crowned Bakelite the King of Plastics. "Your fountain pen, your light switch, your lamp shades and spectacle frames, electrical insulation of all kinds, and if your phone is of the one-piece type, the receiver you hold in your hand, is almost entirely made of Bakelite plastic."

But *was* Bakelite plastic? The term—derived from the Greek *plassein,* which means "mold"—had not yet been universally accepted as a generic term for a class of synthetic and semisynthetic substances known by various names.

Plastics magazine, whose title confirmed the growing popularity of the term, nevertheless awarded a prize to Ottoman Black of Caldwell, New Jersey, for suggesting the name "resoid" as "the best generic name for all molded products." "Molded" had become the preferred euphemism for cheap, tacky plastic.

The National Electrical Manufacturers Association proposed its own candidate: "synthoid." Upon hearing that outrage, Dr. Baekeland used all his remaining—if waning—influence as the reigning "Father of Plastics" to nip that one in the bud.

"The name has no meaning," Baekeland sputtered in an irate letter to *Plastics.* "It barely suggests that the material is made by synthesis." Pointing out with pedantic precision that *-oid* is a Greek suffix meaning "like" or "resembling" (hence anthropoid, asteroid) the aging Ph.D. in chemistry patiently explained that "the ridiculous 'synthoid' could only mean a material resembling synthesis."

"My frank opinion," Baekeland sagely concluded, "is that 'resinoid' would be preferable."

By the late thirties, Bakelite's problems were more serious than the trifling issue of a name for the class of materials to which it belonged. Its latest drawback was that it didn't take colors nearly as well as the new urea-based phenolics like Catalin, whose self-styled "gemlike beauty" had made it the up-to-the-minute casing for a new generation of brightly colored radios.

"Nothing can match [Catalin] in potential gaudiness," *Fortune* digressed with a refined shudder. "There is no law to restrain designers from doing again in all colors of Joseph's coat what they did in simple chromium during the Modernistic days. . . . If Bell Telephone should all of a sudden decree that henceforth all telephones must be pink, Dr. Baekeland would be out of luck."

Seeing the writing on the wall, in 1939, as a second world war approached in Europe, the seventy-five-year-old Baekeland sold out Bakelite General, lock, stock, and barrel, to Union Carbide. By the time he cashed in his chips, Doc had been in semiretirement for years and had been spending more and more of his time at his new winter home in Coconut Grove, Florida, the former estate of William Jennings Bryan.

Doc enjoyed yachting on the inland waterways and making his own wine and beer, a genial habit he acquired during Prohibition. As war loomed, the legendary Father of Plastics spent more time than he could spare attending a global whirl of high-level, high-minded peace conferences, where he would tirelessly expound his pet theory that since science had developed weapons of mass destruction so powerful that they could kill thousands where once they could kill only hundreds, mankind would quickly advance to the general conclusion that war was insane.

As he prepared to meet his maker, Baekeland was feeling more than a tinge of guilt over Bakelite's countless military applications. Three decades later, Stephane Grouff, author of one of the first books about the Manhattan Project, discovered that Bakelite had retained its strategic punch well into the Atomic Age. After sending his national-security-sensitive book to the U.S. Atomic Energy Commission for approval, Grouff received a phone call inviting him down to Washington for a friendly chat. "One of the things they were particularly sensitive about," Grouff recalled, "was anything having to do with Bakelite. . . . So I took certain things about Bakelite out of the manuscript and they sealed this file in my presence and put all their stamps on it, which I signed and then they signed, and then they locked it up." Whatever arcane contribution Bakelite made to the development of the atom bomb remains classified to the present day.

In May 1940, as the war threatened to pull America into its widening mire, a correspondent for *Time* caught up with the "publicity-shy Father of Plastics" on the occasion of his acceptance of a gold medal from Philadelphia's Franklin Institute for his life-long contribution to science. "The gold medal man is less known to the public than the changes his work has brought to the world," *Time* graciously explained. "From toothbrush handles to telephones and false gums to gear wheels Bakelite has become the quintessential modern material."

"God said, 'Let Baekeland be and all was plastics'" had become a popular saying of the time. But as he neared eighty, Doc began to unravel. Always austere and monkish by disposition, in his declining years he took to eating nothing but soup and beans straight out of tin cans and sleeping in a Spartan room, unfurnished except for a plain white cast-iron bedstead. Hoping to do penance for his boundless wealth, he began to affect a tattered old skullcap, which gave him the wizened look of an unworldly scholar.

While in Florida, he favored dressing entirely in white summer costume, from snow white pith helmet to a white linen summer suit, complemented by ice white summer shoes. If he became over-heated, he would casually stroll into his swimming pool and stand blissfully in the deep end, up to his neck, fully clothed, and then stride purposefully back into the sun to rejoin his companions. To anyone present who dared question his sanity, he would confide, with a conspiratorial whisper, "The evaporation keeps you cool."

His son George Washington recalled the day in the early forties when his father (who had once taken long walks with his grandson reeling off Latin names for every plant and animal they saw on the way) arrived at Snug Rock carrying four suitcases, none of which he could figure out how to open. It turned out that he had developed a brilliant system of interlocking keys that, at the last moment, had failed him.

"What he had done," George later recalled, "was to lock the first case, put the key in the second, lock the second, put the key in the third, lock the third, and so on. In that way he didn't have four bulky keys in his pocket but one."

Sadly, Doc had lost his last key. When George finally located a spare inside the house and opened the suitcases, he found all four filled with books. Doc had spent the previous week at one of his interminable peace conferences but had forgotten to pack clothes.

5

Cellophane

"And here's what saves that good aroma!"

PROTECTS CIGARS—AND LETS YOU SEE THEIR QUALITY

EVERY cigar has its loyal fans. Add them all together and you have an enthusiastic bunch of rooters for Cellophane wrapping.

Ten years ago, Du Pont chemists discovered how to make a *moistureproof* transparent wrapping. Wise cigar makers quickly appreciated the value of this new Cellophane protection, which keeps cigars *fresher*, and prevents pocket breakage—yet still lets smokers *see* the color and quality of the leaf.

Today most cigars come in Cellophane transparent wrapping—so you may be sure of *extra freshness* and *extra pleasure* in the smoking. "Cellophane" Division, E. I. du Pont de Nemours & Co., Inc., Empire State Building, New York City.

Cellophane
TRADE MARK
DU PONT

"Cellophane" is the trade-mark of E. I. du Pont de Nemours & Co., Inc.

Courtesy of The Hagley Museum and Library

You're the purple light of a summer night in Spain
You're the National Gallery
You're Garbo's salary
You're cellophane!

—COLE PORTER, *YOU'RE THE TOP* (1934)

I n the fall of 1930, at the tail end of the first year of the Great Depression, the folks at the RJ Reynolds Tobacco Company were finally forced to face facts: Camel, long the dominant cigarette brand in America, was no longer leader of the pack. The year before, upstart Lucky Strike, under the ruthless leadership of George Washington Hill, had boldly unseated Camel from the top spot. Now number three Chesterfield was giving Camel a run for its money. Even Old Gold, until recently hardly a contender, was forcing the American tobacco industry to expand its long-standing Big Three into an all-new Big Four.

Having spent years fretting over Camel's troubles, the top brass at RJ Reynolds at last concluded that something had to be done. So they invited one of New York's highest-powered idea men, O. B. Winters of Erwin, Wasey & Company, down to Winston-Salem, North Carolina, to bat around some ideas on how best to bring the brand back from the brink. If Winters hit a home run, his agency stood to gain the plum Camel account, worth at least $12 million depression-swollen dollars in annual billings. And if Winters struck out . . . he would have no one but himself to blame.

After umpteen hours of marathon brainstorming, not only had Winters failed to hit any home runs, but he had popped up strictly fly balls and fouls. With a frustrated snap of the wrist, he closed his battered leather portfolio, announcing to a weary conference room that much as he had enjoyed their company, it was high time to call it a day. As he made his way toward the parking lot to begin the long drive to his weekend retreat in Maryland's horse country, a Reynolds vice president, hoping to make idle conversation, casually remarked: "You know, we're thinking of coming out in moisture-proof cellophane soon."

Moistureproof cellophane! The dry phrase sounded like music to Winters's ears. With a cry of "Eureka!"—or muffled sounds to that effect—he charged back into the conference room still being vacated by Reynolds executives. After urgently summoning them back into secret session, he waited patiently until his audience had taken their seats and were leaning forward in breathless anticipation before dropping the bombshell of his career.

"Gentlemen," he pronounced solemnly, "I give you . . . "—and with a flamboyant flourish he pulled a fresh pack from his pocket—"the *Humidor Pack*."

Conceptually, the move was sheer genius. What's more, the Reynolds men knew it. Retiring in triumph to his Maryland horse farm to churn out copy over the weekend, Winters spent the next week holed up with three other copywriters at the Drake Hotel in Chicago—where they knew a printer who could be trusted—conjuring up slogans for the proposed campaign.

Back at Reynolds HQ, the home team was so pleased with the results of this crunch effort that they promptly pulled the entire account from its longtime handlers and turned it over to the wizards from New York. As a further sign of confidence, they gave Erwin, Wasey permission to spend the entire previous year's ad budget in the first *two weeks* of the coming campaign.

Sure enough, the Humidor Pack captured the imagination of the American smoker as the first cigarette pack ever guaranteed to "Lock in Freshness."

As a Reynolds stockholders' report proudly put it: "The air-sealed Humidor Pack is capable of keeping in Camels, until they reach the smoker, practically all of that natural freshness and mildness that has always been so carefully safeguarded through our processes of manufacture."

"Freshness" was the prime virtue trumpeted by Camel to compete with Lucky Strike, whose major selling point was that they were "toasted." In the ongoing conflict between fresh and toasted, Du Pont's vaunted research department had played a key role. After buying the rights to cellophane—a French invention—in 1920, Du Pont spent the next seven years and as many millions of dollars trying to make cellophane *moistureproof* instead of just *waterproof*.

Failure to provide an effective vapor barrier had shut cellophane out of the cake business, the bread business, and most importantly, the tobacco business. But once cakes and specialty breads hit the shelves wrapped in Du Pont's brand-new moistureproof cellophane, Du Pont had every reason to expect tobacco to fall neatly into the cellophane fold. But though a handful of cigar makers boldly flirted with the transparent wrapper, cigarettes remained cellophane-free.

Until the morning of June 3, 1930, when a Du Pont vice president in charge of cellophane sales marched into the office of R. D. Long, secretary to Reynolds chief William Reynolds, demanding to see the boss. While this urgent request was being relayed, the chipper fellow calmly offered Long a cigar from a box he had with him, in which each shining specimen was crisply sealed in a sparkling tube of moistureproof cellophane.

The man from Du Pont calmly plucked a cellophane-wrapped cigar from its nest and threw it onto Long's terrazzo marble floor, stomped on it several times, and for good measure rolled it back and forth under his foot before shucking its cellophane wrapper, striking a match, lighting it up, and taking a long, deep draw.

Duly impressed by the fact that after being crushed underfoot, a cellophane-protected cigar could come through, if not smelling exactly like a rose, at least fresh as a daisy, Long heard nothing more about cellophane until late August. One late summer day, out of the blue, Reynolds asked him if he still had that box of cigars the man from Du Pont had dropped off back in June. The meticulous Long reached into a drawer, pulled out the box, and at Reynolds's urgent request, sampled one at random. To his amazement, after three months the darned thing still drew like a dream.

Right then, Reynolds let Long in on a closely held secret: After meeting with the Du Pont representative, he had commissioned a series of tests, conducted at a laboratory in Pittsburgh, designed to evaluate the vapor-sealing capacities of moistureproof cellophane compared to tinfoil, waxed paper, and glassine. The new cellophane had positively withered the competition. This late-summer man-to-man smoke-out was the final taste test, which cellophane had passed with character and distinction. Later that day, Reynolds issued an order for the next batch of Camel cigarettes to be wrapped in what a press release called "a hitherto unknown complete air-seal of moistureproof material."

It took the Package Machinery Company of Pittsburgh until late November to perfect a pack-wrapping machine nimble enough to sheath cigarette packs in cellophane at the going rate of 125 per minute. It took until February 1931 for all existing stocks of unwrapped Camel packs to be depleted. On March 1, Reynolds gave Erwin, Wasey the high sign to kick off their campaign. It commenced at dawn with a barrage of full-page ads in 1,700 daily papers, 2,300 weeklies, and 400 monthlies, bolstered by deep-throated announcers breaking into radio programs with the momentous announcement: "The Camels Are Coming!"

What the Camels were coming *in,* of course, was clearly more to the point. A point driven home by a coast-to-coast contest featuring a then hefty $25,000 first prize to be awarded to that lucky contestant who came up with the most eloquent answer to the following question:

"What significant change has recently been made in the wrapping of the Camel package and what are its advantages to the smoker?"

Nothing could have prepared the Reynolds brass for what happened next. Over a million entries poured into contest headquarters in Winston-Salem before the deadline—midnight, March 5. On the last day, some 5,500 special-delivery and nearly 2,500 registered-mail envelopes arrived, each of which had to be individually signed for by Reynolds officials. A fleet of armored cars transported sacks from post office to contest headquarters, as rows of machines split envelopes open and teams of temporary office girls returned at night to pin the enclosed letters to them, ensuring that addresses wouldn't get lost in the shuffle.

And the top winners were: a Boston milkman, a Brooklyn housewife, a real estate broker from Duluth. All transported to Winston-Salem in a private Pullman car and treated to a lavish dinner, commencing with fruit cocktail, concluding with a ceremonial opening of a number of Camel packs. As the *pièce de résistance*, the smoking of the ritual cigarette.

One fictitious entry, purportedly sent in by a Filipino houseboy, was widely reprinted in newspapers across the country.

Gentleman Dears: Each day I have been almost a total Lucky Smoker until I buy a pack of your kind for little money and wrapped by Selluphane. I like your kind cigarette. Much better since she comes from Selluphane. She has smooth back, this pack and I like to rub hand over glass front. Lucky she is toasted but Selluphane she is not.

Suffice it to say, during those drab depression years, cellophane stood out as a lone spot of brilliance in an otherwise dismal commercial landscape. Cellophane served as the butt of jokes by Eddie Cantor and Al Jolson, as well as the subject of a famous *New Yorker* cartoon depicting a proud, elegantly dressed father being given a first look at his newborn child, peering down at the swaddled bundle being held up for his inspection by a smiling young nurse, and crying out in astonishment: "My word! No cellophane?"

Esquire soon followed suit with a sexy cover cartoon, "The Girl in the Cellophane Dress" perched between two fairly phallic skyscrapers. In late 1932, U.S. Rubber issued its top-of-the-line tire, the Royal Master, packaged in cellophane, to confer an extra dose of glamour and luxury on its flagship product. Cannon Mills wrapped its top-drawer bath towels and sheets in "the glamorous new transparent cellulose sheeting," while a desperate manufacturer of grand pianos, feeling the pinch in the luxury goods market, offered to deliver its best model to the customer's door wrapped in—what else?—the moistureproof cellophane. Even cellophane arrived from the plant sheathed in itself. "To protect it," a company spokesman dryly explained, "from moisture."

America was getting all wrapped up in cellophane. A new humor magazine, *Ballyhoo*, produced a first issue covered in it, as its droll editor, Norman Anthony, showed up at a news conference wearing a cellophane suit. A cellophane bathing suit garnered a predictable amount of publicity, though it was only semitransparent.

Helena Rubinstein offered her new face powder and rouge combination lovingly embraced in cellophane. The powder box was pierced by a transparent cutout section in the lid of the rouge compact, held in place by a cellophane sheathing of the entire lid. "Accept this rouge compact with my compliments," purred an embossed note in Helena's own hand. "From Helena Rubinstein."

One man inclined to take a dim view of these breathtaking events was American Tobacco's George Washington Hill, who had been reduced by Camel's soaring cellophane success to protracted periods of teeth gnashing, long sessions of silent seething, and the occasional berating of long-suffering staffers for not having come up with this cellophane jazz first.

Needless to say, Lucky Strike didn't take long to announce that Luckies would continue to be Toasted, not Fresh. And, by the way, Luckies would soon issue forth in its own cellophane wrapper. But George Washington Hill was too proud an autocrat to humbly announce, "Lucky Strikes—now in moistureproof cellophane too!" Luckies needed some way to turn the tables on Camel, to stand tall in its shimmering cellophane coat while somehow reducing the Humidor Pack to last year's model.

Now it was Lucky Strike's turn to hold an emergency marketing session. At which, one of Hill's packagers declared that the Humidor Pack did have one Achilles' heel. Cellophane's high "edge-resistance" made the thing devilishly difficult to tear open. Trying to tear the top off often resulted in the shredding of the whole pack. Once the vaunted "air-seal" was permanently ruptured, how good a "humidor" pack could it make anyhow?

An excellent point, George Washington Hill allowed. But Hill raised one objection: How did the young man propose to turn cellophane's weakness into Lucky Strike's strength without sabotaging their own cellophane effort?

The poor fellow had no ready answer to that one. So Hill sent his men into the field to come up with one—fast, or it was curtains for the lot of them. Once again, luck shone on Luckies in the form of a go-getter from the American Machine & Foundry Company— chief rival to the firm that made Camel's pack wrappers. He proposed an ingenious solution: Why not insert a series of tiny cuts into the top of the new Lucky pack, permitting the smoker to tear off a corner just large enough to get at the cigarettes? The deep-thinking soul had even dreamed up a perfect name for the doodad: "The Lucky Tab."

Hill was sold. But rather than buy the idea outright, he frugally placed an order with American Machine for special attachments to his current wrapping machines, capable of making the required cuts. Now, Lucky Strike could share in the glory that was cellophane. According to *Fortune,* "Mr. Winters' bright idea was as much a boost for Cellophane as for Camels." With this dazzling new innovation, "If Camels stood out in the popular mind as the giver of Cellophane, to Luckies went all gratitude for perfecting the gift."

Hard to imagine, but the American public was *grateful* for the gift of cellophane. The victory was complete when number four Old Gold, always a touch behind the curve, came out with its new cellophane double pack constructed on the theory that if one layer of cellophane was clearly a good thing, two must be better than one.

By the time 1932 slipped into history, a winner could be declared in the Cellophane Wars: cellophane. As one popular pundit put it, "A yen for transparency had become a national passion." Even when no protective value was needed, when mere visibility was of no obvious material benefit, cellophane's shimmering presence conferred on the most mundane of products a strong dose of sex appeal—though retailers coyly called it "eye appeal."

One major manufacturer of canned foods was shocked to discover that if he wrapped a cluster of cans in a cellophane package, sales soared—even though cellophane added nothing to the package but a touch of disposable gloss.

A national grocery store chain reported a 2,100 percent increase in doughnut sales in *two weeks* after wrapping its doughnuts in cellophane. Market surveys confirmed that housewives felt no compelling urge to buy doughnuts before walking into the store but snapped them up strictly on impulse "because they looked so inviting in transparent packages." Packing pies in paper cases with cellophane inserts caused pie purchases to triple in six months. From celery hearts to cauliflower, things clearly sold better in transparent wrappers.

An extra boon to hard-pressed manufacturers came in the form of the "prepacked unit package." A buyer of golf balls who strolled into a sporting goods store intending to buy just two balls now came home with three—he had no choice because they came wrapped in shiny cellophane three-packs. As *Business Week* cannily observed, "Small items move faster in packages . . . and will be consumed faster, because it is only human nature to be careless where there is plenty." At a time not of plenty, cellophane radiated a strong sense of surplus.

By the time (1934) Cole Porter put cellophane on a par with the National Gallery, Garbo's salary, and a summer night in Spain, the plain-speaking president of Du Pont's cellophane subsidiary, Leonard Yerkes, was getting just a little fed up. To a *Forbes* correspondent seeking to pry from him "cellophane's secret," he snapped: "Madame Consumer wants to see what she buys and she wants it *clean*. That's the secret." Next question.

Still, even Yerkes, in a sentimental moment might have sheepishly admitted: There was something dramatic, even cinematic, about cellophane. In the frothy manner of those escapist musicals of the thirties, cellophane succeeded in wrapping everyday reality in a shiny new coat and tying it off with a glittering bow. Cellophane lent a touch of glamour and gloss to the most mundane objects, enhancing their appeal and allure not by changing their essence but enhancing the context in which they were sold.

Cellophane was alluring, but above all, it was clean. "Cellophane projects an atmosphere of hygiene," a prominent retailer primly maintained. A new generation of consumers in America was becoming acutely germ conscious. Obsessed with catching cooties. Cellophane struck a chord with the new sanitary sensibility. Associated with purity, transparency, and freshness, this French invention became a symbol of the sleek, slick, sterile American way of life. Cellophane stood for today, tomorrow—all those other wrappers, particularly brown paper, seemed tawdry and drab by comparison. So many people in the present all of a sudden seemed to want to distance themselves from their pasts.

A nation of immigrants, which for centuries had swum contentedly in the melting pot, fishing pickles and herring from briny barrels, plucking bread fresh from the oven or bakers' bins, accepting dry goods and drapes direct from the hands of milliners' clerks, buying meat from the butcher and fish from the fishmonger, hand-wrapped in plain paper, now vastly preferred obtaining the same products with a prolonged "shelf life" in hermetically sealed transparent containers.

Being served by other (foreign) hands was out. Self-service was in. A marvelous new invention out of southern California, the supermarket, made everything you bought in it seem as if it had come straight from the factory instead of straight off the farm.

Clarence Birdseye, the frozen-foods pioneer who discovered the secret of "freeze-drying" while observing Eskimos store food on a tour of Alaska, became an early convert to the cellophane cause. "No more kitchen drudgery. Every Birds Eye product is ready for cooking or serving just as you take it from its Cellophane wrapping. Directly from package to stove without cleaning or looking over," promised Birds Eye Frosted Foods—"Sealed and Protected by Cellophane." Clarence even sponsored a "Cellophane Caravan" that toured all the country's major cities, setting up frozen-foods shelves right in the lobbies of elite downtown hotels.

In 1930, Richard Drew, an engineer with the Minnesota Mining and Manufacturing Company (3M), conceived of a way to coat cellophane with pressure-sensitive adhesive. The resulting product was labeled "Scotch" tape in homage to its predecessor, a pressure-sensitive masking tape Drew had developed to make it easier to give twenties cars jazzy two-toned paint jobs. To make the two-inch-wide tape cheaper, adhesive had been applied to the edges only, a cost-cutting measure some customers denounced as unduly stingy. "Take this 'Scotch' tape back to Minnesota," Drew was frequently told, "and tell your 'Scotch' bosses to stick tape all over it." The name obviously stuck. But the new Scotch cellophane tape achieved near universal acceptance only when a 3M sales manager invented a tape dispenser equipped with a built-in cutting blade.

Neatly wrapped in the shimmering cellophane dream, a nightmare lurked, from which cellophane woke one morning with a shimmying shush: Cellophane was tops in the protection racket. As a Du Pont brochure vigilantly proclaimed, "Cellophane protects *you* from dirt's danger." Cellophane thrived in an atmosphere steeped in fear and caution.

> Like two thieves in the night, Dampness and Dryness silently rob many products. Dampness steals the appetizing, crispy crunch of crackers. Dryness steals the tenderness of delicate dainties like marshmallows. Moistureproof cellophane stands guard over foods, day and night. . . .

Even more destructive to the general health was (tum-te-tum-tum) *Dirt*.

> Let the shopping crowds scuffle up dust, let the strange hands grab and paw—Cellophane film is *on the job*, protecting *your* health and *your* pocketbook. Dust, dirt and the germs on inquisitive hands are *kept out* by . . . Cellophane. It keeps out *foreign odors* too.

Even more insidious than dirt . . .

> Dust is the drab veil that smothers clear, lovely color. The dirtiest dirt, ground so fine that you can't even see its true ugliness. In smoky cities it's a powdered mass of acids that eat fabrics. It's a magic carpet for *billions* of germs. . . .

Not to forget the human factor . . .

> Strange hands. Inquisitive hands. Dirty hands. Touching, feeling, examining the things *you* buy in stores. Your sure protection against *hands-across-the-counter* is tough, clear, germ-proof Cellophane. . . .

There was an insidious side to "the silent salesman," the product that, "because what you see is what you get," protected customers not only from dust and dirt but that other despicable d-word of modern life: disappointment. In certain respects, the stuff was a fraud, because while it promoted freshness, it often sealed in staleness. Cellophane promised "extended shelf life" while conveniently glossing over the fact that "fresh" was fast losing its original meaning. Fresh as in "recently cooked or prepared" had been redefined to: "protected from rapid deterioration by an impermeable moistureproof barrier." With cellophane-wrapped perishables being "unit-packaged" like rubber tires, food was free to sit for weeks on the shelf, and taste like rubber tires.

By 1936, cellophane had so effectively penetrated the nooks and crannies of everyday American life that *Fortune,* mightily trying to summarize cellophane, settled for a list of items "now wrapped in it."

> The original documents of the Constitution and the Declaration of Independence . . . violin strings, plated silver, golf tees, suspenders, shirts, hot-water bottles, shoe laces, spark plugs, tapioca, pie, baby carriages, golf balls, pickles, clocks, and a deluxe edition of *The New York Herald Tribune.*

H. L. Mencken's *American Mercury* countered with a contrasting medley:

Hair brushes, alcohol, soap by the cake and by the dozen, sponges, facial tissues, toilet tissue, cough drops, floor mops, doorstops, golf balls, evening bags, emery boards, electric heating pads, men's pajamas, cranberries, sink stoppers, electric razors, lamp shades, hat stands, Mickey Mouse, dress shirts, rice, nail polish, cocktail shakers, noodles, books.

"At the rate this peekaboo covering is being slapped on things," *American Mercury* righteously sniffed, "it will soon be a novelty to find something you can still buy in a state of nature."

Doc Brandenberger's Amazing Dream Coat

One evening in 1904, Dr. Jacques Edwin Brandenberger, a Swiss chemist employed by the French textile concern Blanchisserie et Teniturie de Thaon, happened to witness a regrettable incident in a swanky restaurant in the Vosges region of France. An elderly gentleman seated at the next table accidentally knocked over a bottle of red wine, sending a stain that rapidly spread across the fresh linen tablecloth. A harried waiter came rushing up in a state of great consternation and began sopping, mopping, and whisking the old cloth away before briskly replacing it with a new one, all before the next course was due to begin.

Dr. Brandenberger, then in his mid-thirties, was a specialist in the chemistry of dyeing, printing, and finishing cloth. Observing this scene with the fastidious detachment common to the meticulous Swiss, he couldn't help but experience a powerful wave of revulsion at the sorry state of French tablecloths. He felt moved to ponder the intangible costs—laundry, labor, wear and tear—incurred by restaurants in having to supply fresh linen for every patron, and in some high-class establishments, having to replace soiled cloths between courses simply as a matter of course.

The practical Dr. Brandenberger soon pushed his reverie in a more nuts-and-bolts direction, expanding it into a vision, beatific in its own way, of a snowy white tablecloth covered with some synthetic surface, which when assaulted by grease, food, wine, or other stains, could simply be wiped clean with a damp sponge. No muss, no fuss, no rushing around to prevent damage to the table beneath. Just a swift swipe of a fast cloth, and your table would be clean as a whistle.

The more Brandenberger pondered the problem of manufacturing such a spillproof cloth, the more he realized that this prospective invention could be considered an updated version of the traditional oilcloth that peasant families had spread on kitchen tables for centuries. But with twentieth-century textile technology, the new tablecloth could be a perfect accompaniment to a middle-class, even an upper-class evening meal. It could be printed in brocades, damasks, tulle, crepe de chine, satin, silk—any finish under the sun that the public, in its infinite wisdom, might desire. The most promising coating for such a material, Brandenberger decided on the spot, was viscose. In due course, it would be known as rayon.

Rayon

Paris, France. 1876. The great Louis Pasteur, professor of biochemistry at the École Polytechnique, is asked by the French government to tackle a problem as knotty—if not quite as life-threatening—as the public-health hazards associated with the widespread consumption of contaminated milk. A mysterious blight has been killing off the silkworms of France. No drug or potion seems capable of putting a stop to an epidemic that threatens the survival of a proud industry that has thrived in southern France since the late Middle Ages.

Distilled down to its bare essentials, the ancient Chinese art of sericulture—also known as silk making—amounted to this: When a silkworm chews on the leaves of the mulberry tree, it chemically transforms the ingested cellulose into a clear liquid, which it excretes through a tiny orifice into twin filaments of extraordinarily high tensile strength. Strength has long been an important part of silk's singular allure. But an even more critical component of its cachet is luster, a quality it shares with no other natural fiber.

Just as precious stones depend for their value on the quality of the light they reflect, silk owes its glamour to the way it reflects ambient light. Any curved reflective surface, like a glass rod, gathers in light that strikes it and reflects it back to the eye of the beholder in a bright shining line. Threads of silk resemble miniature glass rods in their capacity to reflect light. Multiply these threads by weaving them into a shimmering surface, and the resulting radiance is nothing less than ethereal.

The idea of breaking silk's natural monopoly by mimicking it in the lab first occurred to the eccentric British natural philosopher,

Dr. Robert Hooke, who in 1664 published his magnum opus, *Micrographia*, a compendium of "useful knowledge" in which he exhorted his fellow scientists to duplicate the "glutinous excrement" of the silkworm—or he would take up the challenge himself. Hooke failed to take up his own gauntlet, but a century later, in 1754, the great French physicist and naturalist René Réaumur picked up the silk route, delicately, with a fine thread and needle.

In *L'Histoire des Insectes,* Réaumur forthrightly declared:

> Silk is just a liquid gum which has been dried. Could we not make silk ourselves with gums and resins? The idea, which might at first sight appear fanciful, is more promising when examined closely. It has been proven that it is possible to make varnishes which possess the essential qualities of silk. If we had threads of varnish, we could make them into fabrics which, by their brilliance and strength, would imitate those of silk and which would equal them in value.

Buried deep in Réaumur's tome was the arresting claim that the webs of several species of spider were finer and stronger than silk. This prompted two enterprising young French milliners to found competing establishments, Cochot's and Louis Bon's, to supply the luxury-craving aristocrats of the ancien régime with what they hoped would soon become all the rage: stockings and gloves woven from spider silk. Cochot's spider silk stockings weighed in at a tantalizing two and three-quarters ounces—the *pair;* Louis Bon's dress gloves tipped the scales at a featherweight three-quarters of an ounce—once again, for the pair.

If the French Revolution hadn't cut customers off from their minute supplies, Bon and Cochot might have thrived in the growing spider silk market. But the decline of the French aristocracy left the British to dominate commerce in worm silk by increasing production and drastically lowering costs. As the Industrial Revolution took off in Britain, forward-looking British silk merchants even dared speculate on silk's long-term potential as a popular luxury.

On a sunny day in June 1846, as René Réaumur was releasing his silken speculations, some sixteen hundred employees of the smart English silk firm of Courtauld, Taylor & Courtauld gathered under a striped silk marquee to celebrate the firm's twenty-first anniversary. Managing partner Peter Alfred Taylor Jr. vowed to the assembled workers that it wouldn't take long to expand the firm's repertory to reach all segments of society.

"Just three hundred years ago," Taylor recalled, "the first pair of silk stockings was worn by Queen Elizabeth. And I am not at all sure that some of our entertainers have not got on silk stockings today!" Personally, the broad-minded Taylor could see "no reason at all why the luxuries as well as the comforts of life should not be brought home to the cottage of the artisan. Which shows that what one age·terms luxuries, may in the next age become by common usage the undisputed right of all."

Admirable sentiments, to be sure. But given the silkworm's pro-clivities, there was no way for Taylor's prophecy to become reality unless some way was found to realize Réaumur's dream of creating a fiber from "threads of varnishes." That challenge was finally taken up, appropriately enough, by a French aristocrat, whose family may well have patronized the establishments of Cochot and Bon. By 1876, when Louis Pasteur was commissioned by the French government to look into the silkworm blight, the French aristocracy was back in full flavor. Pasteur dispatched one of his protégés, Louis Marie Hilaire Bernigaut, Comte de Chardonnet and student at the École Polytechnique, to southern France to conduct preliminary biochemical research on the root causes of the epidemic.

At his family estate in Besançon, not far from the Swiss border, Chardonnet devoted several years to an in-depth investigation of the chemical properties of the leaves, limbs, and trunk of the mulberry tree, from which the silkworm obtains the raw cellulose it later digests into silk. From the dietary needs of the silkworm, he moved on to its productive biochemical action. When examined closely under a powerful microscope, the worm's silk-making apparatus appeared to consist of a single orifice extruding a pair of liquid filaments, which after being exposed to air dried into silk threads.

If Count Chardonnet had not been an enthusiastic amateur photographer, he would probably never have considered liquid collodion a likely candidate for imitating Réaumur's "threads of varnish." But Chardonnet grasped that collodion, pumped in liquid form through fine holes or tubes, might well yield fine filaments resembling silk threads.

By forcing molten collodion through a pair of fine glass capillaries, Chardonnet produced the first reasonable facsimile of a silk monofilament. By 1884, though no closer than he had ever been to finding the root causes of the silkworm blight, he was—as he assured his old friend Pasteur—tackling the problem from a radically different angle. As he advised in a memoir to his fellow members of the French Academy of Science, he was on the verge of producing "an artificial textile material resembling silk."

That announcement, modestly put, sparked a sensation at the academy. They promptly issued him a license to proceed with his research under its official auspices, which helped him obtain a patent on the process. After divulging the details of his operation to the French press, the fourth estate obliged him by publishing a string of glowing accounts predicting the imminent collapse of the centuries-old silk industry. First the silk blight, now threads made from collodion! The resulting furor attracted the attention of a group of French capitalists eager to get in on the ground floor. Capitalized at six million francs, the count's small factory at Besançon was set to produce what its propietor was now modestly calling Chardonnet silk.

Just as Chardonnet's Besançon plant was going on line, the British inventor Joseph Wilson Swan (who had developed a working carbon filament incandescent lamp twenty years before Thomas Edison commercialized it) displayed at an exhibition of inventors in London an intriguing set of doilies crocheted by his wife, fashioned from the very same nitrocellulose filament Swan had recently produced for a new generation of incandescent bulbs. The chauvinistic British press, thrilled to find a local competitor to the imperious French aristocrat, predictably performed handsprings over Mrs. Swan's doilies. A cross-channel battle was, it seemed, shaping up for dominance of the artificial silk trade.

It took the count until 1891 to iron out the kinks in his process, permitting him to debut Chardonnet silk at that year's International Exhibition in Paris. Chardonnet silk turned out to be *the* sensation of the fashionable French Exhibition, with one popular account insisting that the original inspiration for the forced-collodion process had been Chardonnet's close observation of an Italian pasta-making machine.

Chardonnet silk turned out to be sensational in more ways than one. After a fashionable young lady's ball gown, accidentally touched by her escort's cigar, disappeared in a puff of smoke on the ballroom floor, the famously fickle French press—fed by a threatened silk industry—quickly turned from glowing reports to dark accounts of the dangers of "mother-in-law silk." Its only advantage over the real thing now seemed to be the prospect of sending your mother-in-law to an early grave. At the behest of the deeply entrenched silk industry (still reeling from the effects of the blight) the French insurance industry dealt Chardonnet silk a mortal blow by refusing to insure its production on French soil.

In his native country, the count was down for the count. But not, in global terms, out. He hopped the next boat-train to London to meet Joseph Swan, from whom he hoped to license the "denitration" process by which Swan had protected his nitrocellulose lamp filaments from the heat generated by an electrical charge.

Unfortunately for Chardonnet, treating nitrocellulose filament with ammonium sulfide, the essence of Swan's denitration process, reduced flammability, but at a price: the reduction of the fabric's weight by a third. Wet-strength was also cut by half. Toss in an increased propensity to kink and snarl, and about the only resemblance that denitrated Chardonnet silk still had left to the genuine article was its brassy luster. Rather than serve, as its inventor had hoped, as the next staple fiber for the haute-couturiers of Paris, Chardonnet silk was reduced to supplying dry-goods manufacturers with decorative tassels, trimmings, and braids.

A rival cellulose-based synthetic fabric, however, was nearing construction across the Channel. Charles F. Cross and Edward J. Bevan had shared a lab bench at Owens College in Manchester, then as now the bustling center of the British textile industry. While Cross ended up specializing in the problems encountered by textile manufacturers in bleaching jute, Bevan found work advising a local paper company on how best to manage the changeover from cotton and linen rags to wood pulp as raw material.

After graduation, the two men remained in close touch. They quickly came to the conclusion that their separate investigations were linked by a common cellulose thread. As an eloquent Cross would later say of the cellulose molecule: "I stand in its presence in a state of tongue-tied awe. . . . It is not a village but a whole continent, vast and far-reaching."

Cross and Bevan officially incorporated as Cross & Bevan, and immediately joined the ranks of top chemical-consulting firms in Great Britain. After setting up shop in the Jodrell Labs in Kew Gardens on the outskirts of London, their first project (undertaken at the request of a Manchester textile firm) was to study the effects of the "mercerizing" process on cotton cloth. Mercerizing had been developed by British calico printer John Mercer in 1844, who had succeeded in strengthening plain cotton thread and imparting to it a glossy sheen by bathing it in caustic soda.

Cross & Bevan treated a variety of forms of raw cellulose fiber—paper, cotton, flax, hemp—with caustic soda. They then treated the product of that first process with carbon bisulfide. They called the resulting compound cellulose xanthate. After being bathed in a solution of caustic soda and water, cellulose xanthate yielded a thick, syrupy yellow liquid that, for lack of a better name, they called "golden syrup." It wasn't long before the new substance, due to its high viscosity, acquired a label that stuck: "viscose."

Viscose was the most promising cellulose-based material since celluloid. After patenting their formula in 1892, Cross and Bevan formed a syndicate of investors to exploit it commercially. After rounding up an impressive group of backers, including transparent soap magnate Andrew Pears (whose signature product was "so pure you can see right through it") and dynamite magnate Alfred Nobel, they set about perfecting a viscose-based plastic at a small factory in Erith, in Kent. It was to be called "Viscoid."

Clayton Beadle, a partner of Cross and Bevan's, managed to turn out a few minor articles—umbrella handles, combs, and the like—out of Viscoid. But since the firm's senior partners, Cross and Bevan, enjoyed a long history with the textile trade, they soon redirected the energies of the company toward turning out a sturdy, reliable viscose-based fiber.

Unfamiliar with the mechanics of filament spinning, the partners dispatched one of their first batches of viscose to Charles Henry Stearn, Joseph Swan's key associate in the development of the incandescent lamp. Stearn had a factory in Zurich called The Incandescent Lamp Works, where he continued to employ the talented glassblower and mechanical genius who had blown Swan's original light bulbs—Charles Topham Jr.

Topham managed to spin a yarn out of viscose, but it was too weak and brittle, and liable to fray under tension, to compete with natural fibers on the open market. Still, Stearn was sufficiently impressed by viscose's potential to license the spinning rights from Cross & Bevan. In 1899 he formed the Viscose Spinning Syndicate, which attracted an even more glamorous group of investors than the first round of backers. Stearn's lead underwriter was the German prince Henckel von Donnersmarck, a cousin of Bismarck and close friend of the kaiser. From the French side came the Marquis de Beauvoir and Ernest Carnot, son of a former French president and chairman of the powerful Carnot-Chiris industrial group.

But heavy hitters demanded quick results, and as yet Stearn had no product to sell. After experiencing further technical difficulties attempting to spin viscose into a viable yarn, Charles Topham paid a whirlwind visit to Cross and Bevan's viscose plant in Kent, where he conducted a series of experiments designed to increase the strength of the fiber.

Since he happened to be building a new lamp factory at Kew, after a day of frustrating work with viscose spinning, he hopped the late train to London to make a surprise inspection of the plant site. Returning to Erith two days later, Topham was shocked to find the viscose solution he had left behind dramatically transformed. Where before it had dropped straight from its jet into a caustic-soda bath—in which it was supposed to coagulate like an egg white in boiling water—in a chain of distinct beads, the aged viscose now stretched nearly to its breaking point before finally splitting into fragments under the accumulated tension of its own weight. Before it split, the monofilament yarn, when drawn out to its maximum length, was quite strong. By yet another inspired accident, Topham had stumbled upon the value of aging, or "ripening," viscose.

In late 1900, the syndicate's primary investor, Prince Henckel von Donnersmarck, let it be known—in a stern letter to Stearn—that if he didn't see some viscose yarn on his doorstep within a matter of days, he would have no choice but to withdraw his generous support. With a Damoclean sword dangling over his head on a nitrocellulose monofilament wire, Topham took off on a long bike ride in the country to soothe his frazzled nerves. As he peddled anxiously through a deep mud puddle, he happened to glance down at his spinning tire, which flung muddy water across the road in a precise pattern—a perfect semicircle. Peddling with a mounting sense of immiment triumph back to his workshop, Topham sat down and sketched out a rough drawing of a Topham box—a spinning device inspired by the rotary action of his bicycle wheel. The "Topham box," rendered in prototype form, was nothing more than an empty shoe-polish can with a half-inch hole punched into its top at the center. Topham soldered the bottom onto a revolving spindle and attached the spindle to an electric motor capable of spinning the box at the rate of three thousand revolutions per minute.

With hands trembling in anticipation, Topham passed the end of a viscose filament, which he had spun on a bobbin into a ball, over a feed roller until the thread end dropped into the hole in the shoe-polish can. He set the can spinning, which "drew the thread in as fast as it was paid in," Topham later joyfully recalled. "It took twenty minutes to fill the box." Upon opening the box, he found what he was looking for: a piece of cake, otherwise known as "cheese." The viscose cake ran onto its bobbin with great finesse, demonstrating that the threads it contained were strong enough to be spun under high tension.

Topham and Stearn hurriedly spun a dozen skeins of viscose yarn and sent them to Prince Donnersmarck by the next post. The prince had every reason to be pleased by this development because in a matter of months, he was turning out viscose yarn at a rate of twelve thousand pounds per day at his German facility. In 1904, Courtauld's fulfilled its fifty-year-old promise to revolutionize the silk industry in Britain by licensing the spinning rights to viscose from Stearn's Viscose Spinning Syndicate. In America, Cross & Bevan sold similar rights to their good friend Arthur D. Little, the prominent Boston-based chemical consultant, who vainly tried to peddle them in the United States for years without snagging any takers.

Remember the Membrane

Thaon-les-Vosges, France. 1905. Jacques Edwin Brandenberger had been spending the last five years and counting squirting jets of liquid viscose onto cotton cloth, with varying degrees of failure. Lacking a good working knowledge of the use of plasticizers to increase flexibility, his viscose-coated fabrics had turned out glossy but brittle, and most disconcertingly, stiff as boards.

Refusing to give up, and backed by the higher-ups at the sprawling French textile conglomerate Blanchisserie et Teniturie de Thaon, Brandenberger's next step had been to apply a thin film of cast viscose to the cloth surface, which he hoped would stick through the agency of various adhesives. It took another two years to construct a machine capable of casting viscose film in long, continuous transparent sheets. He called these "cellophane"—from the Greek *kellon* ("wood," or cellulose) and *phaino*—"to be seen." As in diaphanous. Transparent.

By 1913, having worked steadily on cellophane for seven years, Brandenberger abandoned his original *raison d'être*—the notion of coating cloth with a liquid viscose film—in order to concentrate on trying to find a ready market for the film itself. The first market he targeted was the cinema industry—cellophane being a less flammable alternative to cellulose nitrate, still banned in France ever since the Bazar de la Charité fire. Le Comptoir des Textiles Artificiels, France's number one viscose producer, decided to back Brandenberger.

In November 1913, on the eve of the First World War, Brandenberger and the Comptoir announced the formation of a new partnership, to be sonorously called "La cellophane." Modestly capitalized at two million French francs, just enough to operate a tiny factory at Thaon-les-Vosges, La cellophane set out to produce nonflammable cinema film. Unfortunately, cellophane film turned out to distort under heat and could not be precisely perforated with sprocket holes.

It's a Wrap

"That," Jacques Brandenberger would later ruefully recall, "was my most difficult time. But when things are really bad, new ideas arise to parry the situation. In the abominable position I was in, I first came to see in my work a new kind of wrapping material."

Wrapping material might have seemed like a comedown from cinema stock, but in fact the potential market was far greater. The only transparent wrapping material available was gelatin, which was at best semitransparent, brittle, unduly affected by moisture, a terrible dust and dirt collector, and easily soiled and dulled when handled. Tinfoil, cellophane's other competitor, was neither as transparent, impermeable to air, nor as light as cellophane.

La cellophane hired a suite of elegant offices on rue St. Cecile, and leased a tiny shop on the exceedingly short rue de la Chaussée d'Antin, just off boulevard Haussman, which extensive research revealed to be the most heavily trafficked commercial street in Paris. Location was critical because cellophane aimed to be the smartest wrapper around. At fifty centimes a square meter, it had to be: Cellophane's first customers kept precious pieces of it securely locked in store safes for fear that some sticky-fingered salesman might make off with it after hours.

The company's first sale, in keeping with its exclusive image, was to Coty, then the most fashionable *parfumeur* in Paris. Coty initially demanded exclusive rights to cellophane as the wrapper for his entire line of perfumes, but Brandenberger persuaded him to settle for an exclusive cellophane sheet, embossed with a silklike surface, which he sold to Coty alone.

The next sale was to Gibbs, producers of the most expensive dentifrice (toothpaste) in France, followed by the baker of the most exclusive gingerbread in France. Cellophane was defining itself as a carriage-trade product, a shimmering jewel in sheet form, the epitome of elegance and glamour in the new commercial art form called packaging.

But Brandenberger never lost sight of the mass market, and had high hopes for penetrating the great American hinterland, starting with chocolates. Cellophane's agent in America, the German-born Franz Euler, set sail for New York armed with a large shipment of cellophane sheets, which he hoped to skillfully wrap around chocolates and other candies from coast to coast. But when Euler opened the package in New York, the sheets had stuck together. Cellophane's bane, excessive moisture, had torpedoed cellophane's American hopes. A few weeks later, the *Lusitania* was torpedoed. Franz Euler was interned as an enemy alien. The time for chocolate bonbons and perfumes was over. Cellophane went to war.

Brandenberger's plant in Thaon-les-Vosges, following the initial German thrust deep into France, was forced to relocate to Bezons, closer to Paris. But there, under enemy fire, having lived a life of frippery and frivolity, cellophane discovered a deadly serious strategic purpose: Its impermeability to poison gas and its tendency not to fog up made it a perfect lens for gas masks.

In January 1920, two years after the armistice, Du Pont—looking to diversify out of the gun- and blasting-powder business—snapped up the American rights to viscose from the giant French viscose producer Le Comptoir des Textiles Artificiels, which also owned a controlling stake in La cellophane. In short order, Du Pont broke the ground on construction for a viscose plant to be located at Buffalo, New York, on the shores of Lake Erie, designed to be an exact duplicate of the Comptoir's vast viscose facility in France.

After sealing the deal on viscose, the Comptoir dropped a casual line to their counterparts at Du Pont, advising them that they owned the rights to "another promising cellulose product in which your company might possibly be interested."

The head of Du Pont's mission to France, executive committee member William C. "Colonel Bill" Spruance, knew something about cellophane. Twenty years before, Spruance had accompanied Arthur D. Little to England on an inspection tour of Cross & Bevan's viscose plant at Erith, and had been privy to a pilot project in which Andrew Pears wrapped a small number of bars of Pears Transparent Soap in transparent viscose, years before Brandenberger ever dreamed of cellophane as a packaging material.

Du Pont was interested in cellophane, and after quickly coming to an agreement with the Comptoir, elected to build a cellophane plant right next to its viscose plant in Buffalo. That way, if the cellophane business didn't work out, the factory could—with the turn of just a few screws—be converted into a viscose-fiber factory without suffering any great loss.

On April 4, 1924, Jacques Edwin Brandenberger, in shirtsleeves, pulled the first cellophane sheet hot off the roller. By all accounts, confusion reigned as French experts screamed at their Yankee counterparts in exasperation, trying to make themselves understood through a lone interpreter.

Compared to marketing cellophane, making it was a lead-pipe cinch. Du Pont had no patience for the carriage trade. "If we can't sell this stuff by the carload," cellophane marketing director Oliver Benz told his colleagues on his first day on the cellophane detail, "I'm not interested in it." At close to five dollars a pound, cellophane was a tough sell.

Brandenberger had been right about one thing: Chocolate was

the perfect wedge to break into the food-wrapping market. Cellophane's first conquest in America was Whitman's Chocolates of Philadelphia, which agreed to test it out after Benz offered to hand-wrap and deliver 40 percent of the next production cycle as a trial run. Whitman's medium-priced candies sold well in cellophane, but Yerkes, chairman of the new rayon division (which included cellophane), battled with his superiors for a massive price cut (down to $1.75 a pound), for cellophane was impossible to move in bulk.

The price cut gave take-no-prisoners cellophane salesman "Kick" Jorgenson the nerve to stroll into the office of a Cleveland cookie maker named Wolf, known to be one of the hardest nuts to crack in the business. As Jorgenson delivered his standard cellophane spiel, Wolf turned his back, cutting Jorgenson off in midsentence, stood up, and left the room. Jorgenson retaliated by picking up some cookies lying loose on Wolf's desk, neatly wrapping them in cellophane before leaving. When he stopped by Wolf's a month later, he had his first order.

Cellophane was promoted by Du Pont as "thin as tissue but hard to tear, like paper but not paper, transparent as glass but not glass—a non-fragile water-proof product with a singularly wide range of uses." But its "singularly wide range of uses" was in fact limited by poor performance as a vapor barrier. When a hard-won customer, Ward's bakery, agreed to wrap a line of premium-priced cakes in cellophane in late 1925, the cakes rapidly turned into rocks. That was the last straw: Until it could be made moisture-proof, as far as Du Pont was concerned, cellophane would stay in the doghouse.

In desperation, Yerkes had a word with Du Pont Chemical director Ernest Benger, who in turn requested $6,000 in new laboratory equipment and enough money for a new hire to solve cellophane's moisture problem once and for all. Benger chose William Hale Charch, a twenty-seven-year-old chemistry Ph.D. recently laid off by General Motors—which at the time was almost entirely owned by Du Pont—as his man on the beat. When he joined Du Pont, Charch had been out of graduate school for only eighteen months.

Charch tried rubber, latex, and other waterproof gums before settling on nitrocellulose as a coating for cellophane. His research was aided by Du Pont's recent success with Duco car finishes, based on "quick-dry" cellulose lacquers that had revolutionized the automobile industry by accelerating the painting process and vastly expanding the available color range.

Charch soon discovered that a composition of nitrocellulose and wax increased cellophane's moisture resistance a hundredfold. Still, they encountered innumerable obstacles involving flexibility and adhesion before finally perfecting a complex four-part system, which featured a nitrocellulose layer, a wax moistureproof barrier, a plasticizer, and a blending agent. Charch's coating added a mere 10 percent to the weight of the film, while still managing to remain under one ten-thousandth of an inch thick. To moistureproof an acre of cellophane, Charch needed a mere eight ounces of nitrocellulose wax.

In June 1926, the first five-hundred-pound batch of Du Pont's moistureproof cellophane rolled hot off the presses in Buffalo. Four years later, the Humidor Pack and the Lucky Tab made cellophane famous.

Rayon's Rebound

That year (1926) the Paris couture houses decreed that luster was out. *Vogue* delivered the coup de grâce when its editor-in-chief starkly stated that "the sheen on women's clothes would soon be as obsolete as the sheen on her nose."

That was certainly bad news for rayon—the new name for viscose recently approved by the National Retail Dry Goods Association, a department and drapery store trade group, in response to complaints that, as one distressed dressmaker put it, "Calling viscose Artificial Silk is tantamount to calling a steel beam 'artificial timber,' or a brick 'artificial stone.'" Acutely sensitive to these concerns, and aware that the name "viscose" was the furthest thing from sexy, the NRDGA launched a nationwide contest to find a new name for viscose. The contest had only one condition: The new name could not include "silk."

Du Pont's earlier trademark, Fibersilk, was clearly out as a generic term. New names suggested by telephone, letter, and telegram included "filatex," "glis," "glistra," and the ever popular "klis"—"silk" spelt backwards. An aging Thomas Edison, always available to put in his two cents, suggested "plant silk" to distinguish it from "worm silk," the ban on "silk" having obviously slipped his mind. And the winner was . . . rayon! Proposed by Kenneth Lord of Galey & Lord, an old-line textile firm, "rayon" was said to derive from the French for "shine"—a reference to viscose's main selling point: its silklike luster.

But by the spring of 1926, with moistureproof cellophane just coming off the line, Du Pont's research department under Benger was forced to swing into action. Its mission, should it choose to accept it, was to deluster rayon. Before the scientists in Wilmington could even reach for their beakers, word reached them that another company had discovered an easy, cost-effective way to do the job. Apparently, adding trace amounts of titanium oxide—white paint pigment—wiped the sheen off rayon yarn in a jiffy, and at next to no cost.

Delustered rayon turned out to be the best thing that ever happened to viscose since it gained a new name. Rayon became gorgeously, tastefully dull. The French couturiers, who had always despised synthetics, simply adored it. The French textile mills began promoting it. Elsa Schiaparelli, a young avant-garde designer and daughter of a prominent Italian astronomer, had been turning up her shapely nose at the run-of-the-mill fabrics being shown her by a famous French textile house when she demanded to see the "howlers"—those experimental novelties discarded as too extreme or impractical.

Reluctantly, they pulled out a few stray bolts of delustered rayon, from which she selected a heavy serpentine crepe and turned it into a form-fitting evening gown. It lent even the clumsiest wearer a touch of slinky Hollywood class. She created a cape for her winter collection that she claimed was out of "glass." She glamorously referred to it as Rodophone, but it was cellophane, darling, and to hell with it.

"Rayon," proclaimed Saks–Fifth Avenue in 1930, "is like the times we live in! Gay, colorful, luminous. It is so pliable to work with and so luxurious in appearance. It launders to perfection. . . . "

It had only one drawback: It didn't dry-clean well. When *Vogue*'s Diana Vreeland sent a slinky Schiaparelli dress to her local dry cleaners, she received a call from the corner from a voice in distress, advising her that it had disappeared.

"You mean you *lost* it?" Vreeland cried out, horrified.

"No ma'am," came the sheepish reply. "It dissolved in the dry-cleaning fluid."

Not for nothing did Schiaparelli give her new fabric a space-age name: Cosmic. As rayon, viscose, or what have you, cellophane had become cosmic. Fashion arbiters ascribed rayon's success to the changing contemporary silhouette "from the short, straight chemise frock to a more form-fitting, drapey, curved line, conforming more closely to the natural line of the figure." The supremacy of the "sculptured" silhouette was a by-product of "the glorification of glamour in motion pictures." Illustrating that hypothesis, the 1935 Hollywood version of *A Midsummer Night's Dream* featured dozens of fair maidens with shapely legs and breasts gamboling through sylvan woods draped in nothing but sheer, shredded cellophane.

Hemlines helped. As they rose smartly from six inches above the ground in 1919 to twelve inches above in 1920 to a scandalous eighteen inches in 1925, hose became an increasingly critical fashion statement. Hollywood helped by draping its virginal heroines in cotton gingham and linen, its vampy villainesses in scandalous, slinky synthetics. No modern woman wanted to be a virginal heroine. Sales of rayon underwear, silky sexy but cheaper, shot up fivefold between 1925 and 1928. By the mid-thirties, an estimated 85 percent of the dresses sold in America contained at least a little rayon.

From its debut in Paris as a glamorous wrapper for Coty, two decades later it had smartly moved both upmarket and downmarket, from Elsa Schiaparelli nightgowns to Whitman's chocolates and Camel cigarettes. When Virgil Thomson and Gertrude Stein produced their avant-garde play *Four Saints in Three Acts,* it featured an all-cellophane decor by Florine Stettheimer, a little-known painter destined to achieve fame a half-century later when New York's Whitney Museum posthumously organized a one-woman show in her honor.

At the Ballet Morkin, a group of ghostly young women, shades who had died unseemly deaths as the result of unrequited love, expressed their passionately ethereal nature by draping themselves from head to toe in transparent cellophane bands flowing down from their arms and hair. Paramount's 1937 musical *The Big Broadcast* featured a variety number starring Jack Benny, George

Burns, and Gracie Allen backed by dancing girls in transparent cellophane top hats.

Despite its common association with frivolity and froth, cellophane had by then acquired a few deadly serious uses. It made the ideal semipermeable membrane for kidney dialysis, and a transparent surgical dressing for wounds—enabling physicians to gain a clearer view of the healing process.

In 1931, the year of the Humidor Pack and Scotch tape, an animal behavior researcher at Brown University discovered that cellophane cigarette wrappers made the perfect semipermeable membrane for a pneumatic tambour designed to record the breathing patterns of white rats. Scientists at a Connecticut Agricultural Station used colored cellophane sheets to "determine the wave length of light responsible for producing sun-red pigment in maize plants." At Johns Hopkins Hospital, cellophane was found to surpass glass as a transparent material for transmitting UV rays. Wearers of prosthetic limbs discovered that a cellophane sheet slipped in the joint between the artificial limb and the skin reduced chafing more than any other material, natural or synthetic. This was as it should be, since cellophane increasingly bridged the widening gap between the natural and the synthetic worlds.

The daring Professor C. M. Mackay of the Lab for Animal Nutrition at Cornell University in Ithaca, New York, reported that over the summer of 1931 he had "fed college students on a diet of cellophane, producing no serious adverse physiological or emotional disturbances." His stated research purpose in feeding his students cellophane was "to create diets allowing fat people desiring to reduce to ingest bulk without calories."

In 1940, cellophane crowned its ethereal dominance of the depression decade by placing close to the top in a nationwide poll designed to determine "the most beautiful words in the English language."

Cellophane placed third—beaten by "mother" and "memory."

6

Nylon

Courtesy of The Hagley Museum and Library

In olden days a glimpse of stocking
Was looked on as something shocking,
Now heaven knows, anything goes!
— Cole Porter, *Anything Goes* (1934)

S hirley Temple couldn't make it in person. But her happy little disembodied voice could, piped in by phone from a Hollywood film set, where the cameras briefly ground to a halt so she could bid the crowd in New York a big warm hello: "Gee, I wish I could be with you at the fair today! Have a good time!"

"Marvels of Tomorrow Unveiled at Fair Ground As Forum Ends," saluted the headline in the *New York Herald Tribune,* official sponsor of the Eighth Annual Herald Tribune Forum on Current Problems, an event that drew over three thousand women's club members to an auditorium in Flushing, Queens, constructed on the site of the 1939 New York World's Fair. There, they listened appreciatively to a series of prominent speakers holding forth on the organizing theme of the upcoming fair: "Entering the World of Tomorrow."

Following such illustrious speakers as Eleanor Roosevelt, "master builder" Robert Moses, social activist Dorothy Thompson, and Republican presidential candidate Wendell Willkie, Con Edison chairman Floyd Carlisle confidently predicted that "New Yorkers will walk the streets at night illuminated to the brightness of daylight, flooded with changing colors, to the accompaniment of symphonic music, synthetically produced by the blending of electronic waves of various frequencies."

In a similarly utopian vein, Lewis Waters, head of research for General Foods, painted a cozy domestic scene in which "New Yorkers of tomorrow will eat more attractively colored instant foods, cleverly packaged in containers so elegant they can be set directly on the table."

This gee-whiz tone was admirably sustained by Dr. Charles M. Stine, vice president of E. I. du Pont de Nemours, who mounted the podium to share with the room a few upbeat thoughts about plastic. There was, apparently, a bright future in it.

> Plastics are to be regarded as *new* materials, rather than as substitutes for *old* materials. Look about your automobile and your home and you will find a variety of fabricated articles made from plastics.
>
> For example: beautiful and durable toiletware, toothbrush handles which have entirely replaced the bone handles of bygone days, costume jewelry, unbreakable tableware, electrical appliances, lighting equipment, buttons and buckles, scuffless shoe heels, motion picture film, radio cabinets, steering wheels, instrument panels, shatterproof glass . . .

Before this carefully chosen group of professional women— every one of whom could be counted on to be wearing beneath her tailored skirt a pair of full-fashioned silk hose—Stine raised his voice a few carefully calibrated decibels to announce the impending debut of the female leg-packaging material of tomorrow.

"To this audience, I am making the first announcement of a brand-new chemical textile fiber. . . . "

The room lapsed into a stunned silence. Rumors of just such a breakthrough had surfaced in the business and trade press in recent months.

> The first man-made organic textile fabric prepared entirely from new materials from the mineral kingdom.
>
> I refer to the fiber produced from nylon. . . .
>
> Though wholly fabricated from such common raw materials as coal, water, and air, nylon can be fashioned from filaments as strong as steel, as fine as a spider's web, yet more elastic than any of the common natural fibers and possessing a beautiful luster.

At those ringing words "strong as steel," the audience burst into a spontaneous round of applause, clearly under the false impression that stockings "strong as steel" would not run. Not in the World of Tomorrow! Just as foods would be packaged in plastic containers so fine they could be set directly on the table, just as cellophane would permanently protect us from dirt's danger, now nylon would attractively and durably package a woman's legs in a sturdy synthetic sheath, combining yesterday's elegance with tomorrow's convenience.

Five months earlier (July 1938) textile reporter Stephen S. Marks had broken the story in New York's *Daily News Record* of the final achievement of the dream of Hooke in 1664, Réaumur in 1754, Chardonnet in 1884, and Cross & Bevan in 1892—to create a synthetic silk virtually indistinguishable from the real thing. If nylon proved to be half what Du Pont had cracked it up to be, there was definitely a bright future in it.

ENTIRELY NEW SYNTHETIC YARN FOR WOMEN'S
HOSE DEVELOPED WILL BE KNOWN AS FIBER 66

"There are reports that it is a protein substance bearing a close chemical resemblance to silk," Marks tentatively asserted. "However, others seem to think that it is one of the new complex chemical compounds such as many of the new plastics. . . . " Being no chemist, how was Marks to know that Fiber 66 was in fact all the above?

The natural fibers nylon aimed to displace were pretty much sitting ducks: Silk was expensive, and not very durable—a pair of superfine single-thread silk stockings couldn't be counted upon to outlast an entire evening's gaiety without running. Wool was scratchy and dowdy. Cotton was sturdy and serviceable, but hardly the last word in elegance. Rayon was neither cheap, strong, smooth, nor sheer enough to seriously compete with "the queen of textiles."

Sheerness—the sheerer the better—had taken on major social and cultural significance since the end of the First World War, when hemlines began their inexorable rise. At the turn of the century, *Vogue* had decreed an embroidered, bejeweled black silk stocking (circa 1902) to have attained "*almost* the limit of perfection."

But those stockings were opaque leg warmers, veiled by street-sweeping skirts—achieving at their sauciest a coy peekaboo bravado. In olden days a glimpse of stocking *had* been looked on as something shocking. But by 1915, cut-and-sewn glove-silk stockings, in all black, all gray, or gray-and-white or black-and-white stripes, were meant to be seen only by the most private eyes. By 1920, with hemlines dramatically up, an even more shocking expanse of leg was expected to match a woman's shoes—in shades of black, white, brown, or gray. A woman with a yen for something more dashing could buy a "luxury" twelve-thread stocking—all silk from hem to toe—for four dollars. At a time when working men were paid on average five dollars a day, "silk-stocking district" was more than just a turn of phrase.

By 1925, the height of the flapper era, neutral colors had come to the fore: dirty champagne, beige, taupe, gunmetal. By 1930, sheer silk stockings, mainly chiffons, executed in subtle colors were all the fashion in the "delustered" age. Crepe twist silk stockings, in up to twenty different shades, were the final word in elegance. By 1938, American women were buying over five hundred million pairs of silk hose a year, worth an estimated $700 million to Japanese silk producers alone. Working women (of whom there were more every year) could be counted on to run through thirty-six pairs of silk stockings in a year.

Nylon was not only a potentially mortal threat to the ancient silk industry, but a lifesaver for Western toothbrush manufacturers, cut off since 1937 from their customary supplies of Chinese superfine swine bristles by Japan's brutal invasion of Manchuria. Two days after Stine's announcement, a double-page ad appeared in the *Saturday Evening Post*.

NOW A TOOTHBRUSH WITHOUT BRISTLES!
The New Dr. West's Miracle-Tuft
Ends Animal Bristle Troubles Forever!
YES, IT HAS NO BRISTLES!
CANNOT SHED! CANNOT GROW SOGGY!
CANNOT FAIL TO CLEAN!

Du Pont's newest miracle fiber, identified only as "Exton," was guaranteed not to shed, break, pull out in your mouth, go soggy and limp when wet, or otherwise pose the various and sundry "animal bristle troubles" that had bedeviled devoted toothbrushers ever since the 1780 invention of the wooden-handled, boar-bristled toothbrush by William Addis of Clerkenwell, England.

Nylon had come to the rescue in the guise of a mysterious stealth fiber known as Exton, the alias under which a canny Du Pont had elected to release a cruder, thicker version of nylon (which could be produced in tubes the thickness of a man's arm). In this relatively innocuous application, under an assumed name, nylon was saved from having to go up against silk until every conceivable kink had been worked out. In 1938, months before Stine's announcement, 15 percent of higher-priced toothbrushes were being sold with nylon—"Exton"—bristles.

Stine's poetic claim that nylon was "wholly fabricated from such common raw materials as coal, water, and air" was taken literally by some inspired cartoonists, one of whom portrayed a woman's darning basket containing a lump of coal, an electric fan, and a bottle of water. Another showed a man upbraiding his wife: "You gotta quit trying to sneak those nylon stockings in with the coal bills." Still another depicted an irate woman trying to return a pair of nylon stockings with a run in them: "If nylon is made from coal, water and air, someone put too much air in this pair!"

The *Science News Letter* took the fundamentals of organic chemistry to macabre extremes, pointing out that since a primary ingredient in nylon is coal tar, the origin of "thousands of organic compounds ranging from perfumes which nature never knew to explosives and dyes and even organic compounds in the body itself," there was no reason why nylon couldn't be made from "a human corpse . . . since a basic ingredient of the fiber is the evil-smelling poisonous compound *cadaverine*."

Such morbid thoughts might have been more understandable if coming from a Japanese silk merchant, most of whom felt unfairly targeted by nylon's aggressive sneak attack—a pre–Pearl Harbor role reversal. In those preelectronics days, silk was Japan's number one export commodity. Of Japan's nearly $150 million in sales to the United States in the first eight months of 1939, more than 65 percent had been silk. "New Synthetic Fiber May Smash Jap Monopoly," reported the *Philadelphia Record* the day after Stine's announcement. "Nylon may end the Japanese monopoly on silk, which nets the Japanese more than $70,000,000 annually from American hosiery makers alone."

By invading Manchuria and committing the brutal Rape of Nanking, Japan had alienated millions of high-minded customers in the West. "I do not intend," one irate young woman wrote to Du Pont, "to pay for a single silk stocking until the Japanese get out of China." Nylon's success posed a serious threat to the Japanese war machine, which had been counting on high foreign silk sales to finance massive imports of munitions. All of which led to documented distress among "incensed Japanese silk men, among whom it is common knowledge that the name had been adopted for the express purpose of ridiculing Japan's Agriculture and Forest Ministry," according to the militaristic *Japan Times Weekly & Transpacific*. Nylon spelled backwards, you see, is 'nolyn,' which sounds like 'norin,' which means 'agriculture and forestry' in Japanese. This powerful government ministry, bureaucratic overlord of Japanese silk production, was "the primary target of the president of the Du Pont company," the *Japan Times* explained. "Mr. W. S. Carpenter hates the Japanese, and has long felt that importing large quantities of silk from Japan is an abominable practice which must be thwarted at all costs. Naturally, this prompted him to look around for an ideal substitute, which will deal a death blow to Japanese silk."

Nylon, according to this defensive view, stood not merely for "agriculture and forestry" spelled backwards but for something far, far more nefarious. "By taking the first letter from every word of this spiteful sentence, the Du Pont president himself has personally selected the name "nylon" to mean:

"Now You Lousy Old Nipponese!"

Hoping to soothe ruffled feathers on the Pacific Rim, Du Pont vice president Leonard Yerkes dashed off an appeasing note to the Fiber Supply and Demand Adjustment Association of Japan, disclaiming all sinister intent.

> I adopted the word "nylon" because it was plain, easy to pronounce and easy to memorize. Please rest assured . . . when we state that the word 'nylon' positively contains no malicious implications.

The name had in fact been the end product of an exhaustive linguistic search for the right handle for a product that had devoured twelve years and an estimated twenty-seven million depression-era dollars in development funds. In January 1936, with "Fiber 66" about to go on-line, Benjamin May, vice president of Du Pont rayon, sent a memo to Yerkes, broaching this delicate subject.

> In connection with the Experimental Station project for a new synthetic fiber, sometimes referred to as Rayon #66, I wish to ask that consideration be given . . . to whether it is wise or unwise to denominate it as "rayon."
> Fiber #66 has many outstanding properties superior to what to date has been comprehended under the word "rayon"—it may be so distinctly different and superior that to call it something else might be better.

Yerkes could not have agreed more. He asked Dr. William Hale Charch, developer of moistureproof cellophane, to evaluate a few trial names suggested by Dr. Ernest K. Gladding, one of Charch's chief scientists and a future manager of the nylon division.

Gladding's personal favorite was "Duparooh"—an acronym of "Du Pont pulls a rabbit out of the hat." As an alternative, he suggested "Dupron," short for "Du Pont pulls rabbit out of nitrogen, nature, nozzle, or naphtha." Fine, but no cigar. Personally, Du Pont president Lammot du Pont approved of "Delawear." He even came up with a catchy slogan to go with it: "Like the first state, it's the first synthetic textile!"

As a poor second best, Mr. Du Pont had good feelings about "Neosheen." Vice President Carpenter tossed "Duponese," "Pontella," and "Lustrol" into the ring. All found backers, but no takers. The issue was clearly too significant to leave up to any one man. A "Name for Fiber 66 Committee" was hastily convened, composed of Gladding, Yerkes, and May. Dr. Preston Hoff, a colleague of Gladding's in the rayon department, served as official "Coordinator of Suggestions."

Suggestions included a sentimental favorite—"Wacara"—a contraction of Wallace Carothers, nylon's inventor, recently deceased. High-minded, but a commercial nonstarter. Gradually, a consensus formed that a name incorporating "nu" (newspeak for new") might be a good idea. The team quickly ran through variants—"nuyarn," "nuyar" and "nuya"—before settling, briefly, on "nuray." A clear nod toward rayon, and a strong favorite of Yerkes, an old rayon hand.

But nuray's selection, if approved, would have been a grave insult to Fiber 66, which was of far greater scientific and cultural significance than rayon. Apart from the coincidence that they both shared some superficial properties with silk, rayon and nylon could not have been less alike. Rayon and cellulose acetate, the original synthetic fibers, are artificial modifications of the cellulose molecule. But Fiber 66 was a pure test-tube baby, composed of molecules with no counterparts in nature. One hundred percent artificial and proud of it, any name that gave the impression that it was ashamed of its synthetic origins would have been nixed on the spot.

"Exton," test-marketed as Dr. West's Miracle-Tuft, had its adherents, but apart from the fact that it conflicted with an existing trademark, nobody much liked it anyway. The inevitable "Klis," which had also been proposed for viscose, was just too silly to be taken seriously. After actively considering some 350 more candidates, including—

Amidarn, Amiray, Amoray, Amido Silk, Artex, Ceron, Duamex, Dualin, Dusilk, Dextra, Eidaise, Eidu, Elastra, Filosel, Linex, Lamiam, Longduray, Lastica, Lastrapon, Mertal, Mouran, Moursheen, Nepon, Novasilk, Nyzara, Pantex, Pontex, Poya, Ramex, Ramisil, Rayamide, Supralin, Silpon, Syntex, Silf, Silmon, Siltra, Silkex, Silpnex, Tane, Tensheer, Terikon, Wiralene

—Gladding zeroed in on "norun." It was catchy and to the point. But in the light of tests showing that Fiber 66 *did* run, to adopt a name that promised otherwise would have been a public-relations fiasco. In mounting frustration, Gladding tried flipping "norun" into "nuron." This had the drawback of sounding too much "like a nerve tonic," given its unfortunate resemblance to "neuron." Not to mention "moron."

Out went the offending *r*, exchanged for an *l*, yielding "nulon." Which hung in there until it occurred to Gladding that ads referring to "new nulon" might sound redundant. Not to mention the fact that a casual listener might easily misspell it as "newlon." Gladding struck out the *u* and inserted an *i*, yielding "nilon." But even that could still be pronounced three different ways: "nee-lon," "nil-lon," and "nigh-lon." After a marathon meeting at which the committee voted unanimously for pronunciation number three, the spelling was amended to "nylon" to prevent confusion.

September 1, 1938
To: Executive Committee

From: L. A. Yerkes, General Manager, Rayon Department.

The name Nylon has been selected . . . we feel it is the best suggested and should be satisfactory. . . . Permission to register as trade-marks other names denoting products made from NYLON will be requested of your Committee from time to time.

Following Stine's announcement, a Milwaukee man wrote to Du Pont, taking it to task for failing to adopt a "bolder, more intellectually honest name: 'American Cilc.'" Within Du Pont, of the two employees named Nylen (a common Scandinavian name) one actually ended up peddling nylon. And since "nylon" sounded like "Trylon," the tapering tower symbolizing the New York World's Fair, a fair proportion of the forty-five million visitors to the fair simply assumed that nylon was inspired by Trylon. Its discovery was, after all, announced on the fairgrounds. Still others were arbitrarily convinced that nylon was a contraction of "New York" and "London"—a fusion of "NY" and "Lon."

In America, one lone holdout was as incensed as any Japanese silk man at the enormity of the new name.

Mr. W. S. Carpenter, President
E. I. du Pont de Nemours & Co.
Wilmington Del

Dear Mr. Carpenter:

As you will note, our name goes back quite a few years in American history, even if its wearers were clerks, hod-carriers, or something equally prosaic.

But now, alas! My wife goes into a store to make a purchase and she tells the clerk her name. The clerk laughs roguishly and exclaims,

"It's *almost* Nylon isn't it!"

It was amusing at first, but now it has become alarming. Even our old friends don't know us anymore when they hear people hailing us as "Nylon."

They all end up saying sweetly (as if somehow they realized the mischievousness of it all) "Thank you Mr. (or Mrs.) Nylon."

Of course we feel so absolutely helpless about it all. . . . We only wish the corporation had been a little more considerate in its choice of name or at least notified us in advance. Now our name will always be "Nylon." The sad part of it is that our children are still young. We can only hope that they won't allow it to effect [sic] their careers.

Mr. John L. Naylon
6136 Locust St.
Kansas City, Mo

Dear Mr. Naylon:

I can't help but feel that in spite of all you say you must be exaggerating to some extent your situation. There are many people named "Cotton," there are many people named "Woolley" and there are undoubtedly people with names similar to silk, flax, linen, etc.

We appreciate your writing to us as it has given us a point of view which we didn't appreciate before.

Yours Very Truly,
W. S. Carpenter

Du Pont's challenge became not promoting nylon as a substitute for silk but of dampening unrealistic expectations that it would surpass silk in every respect. That catchy phrase "strong as steel," though technically accurate, was clearly a culprit. One hosiery buyer advised Du Pont of his mounting apprehension that "female hopes were running so impossibly high that N-Day would inevitably bring disillusionment." Another buyer insisted he would "rather see women prepared for a superior stocking than a miracle."

In the months preceding nylon's debut on the hosiery-counter shelves, Du Pont's publicity department worked overtime to pursue, as one report put it, "a determined policy of understatement, in order to disarm exaggerated reports that nylon stockings are runproof and that two pairs will last a woman a year." When an early press release asserted that nylon was "one-and-a-half times stronger than silk," word came down from on high that though that might well be the case, future press releases should be amended to make a more modest claim: "Nylon will wear, on average, as well as silk hose."

To avoid a mistake they had made with rayon, when Du Pont called viscose fiber "Fibersilk" when it was plainly inferior to the real thing, Du Pont insisted that the name, which smacked of no mere imitation, be stamped or knitted into every pair of nylon hosiery, conveying pride in the product and an unswerving conviction that nylon stood second to none—not even its natural competitor.

Still, as N-day approached, a flurry of rumors remained to be squelched if nylon stockings were not to be oversold. Among the most rampant:

1. Nylon stockings were runproof.
2. Two pair would last the average working woman a year.
3. Nylon threads were impervious to razor blades and nail files.
4. You couldn't burn a hole in a nylon stocking with a lighted cigarette.
5. To put a run in it, a girl had to use an acetylene torch.

A story circulated by a threatened silk industry claimed that a Vassar girl who accidentally touched a lighted cigarette to a pair of experimental nylon hose had been burned to death. In fact, nylon is virtually fireproof. The most outlandish rumor, but the hardest to kill, was that Du Pont was deliberately debasing the quality of nylon at the request of hosiery manufacturers, who had no great interest in producing an invulnerable stocking.

As the *Milwaukee Sentinel* pointed out, nylon's durability might even have its drawbacks.

> Women may cheer but we doubt women will in the long run—no pun intended—thank American industry for nylon. The ladies may squawk about runs in their stockings, but now they'll howl about 'having to wear the same old thing every day.'

At the Fair (1939)

Du Pont's Wonder World of Chemistry, a towering steel skeleton resembling an industrial-strength Eiffel Tower, had been designed to reflect the skeletal structural character of carbon-based molecules—the fundamental basis of organic chemistry. Milling throngs, inside and out, were treated to endless displays of nylon hose, toothbrush bristles, and fishing line, nylon being the wonder surpassing all wonders.

In a floodlit courtyard, an illuminated fountain composed of thousands of nylon filaments, animated by air currents and spotlighted at night, set the stage for the world premiere of *Nylon: This Chemical Marvel*. A knitting machine knit nylon stockings right before your very eyes. A pair of automatic mechanical hands stretched a nylon stocking to its longest length and snapped it back, day and night, around the clock, proving to skeptical visitors that nylon was surely the strongest, most elastic fiber ever invented.

Du Pont's floor show—better known as a "leg show"—was judged "sexiest corporate show at the fair" by one impassioned observer, who favorably compared the shapely charms of Miss Chemistry—who reclined on a plastic podium demonstrating her shapely gams sheathed in supersheer nylon stockings to the approval of mostly male crowds—to the out-and-out striptease extravaganzas being touted on the vulgar "Midway"—the honky-tonk concourse.

Pictures of lovely lasses pulling back long flared skirts to reveal long nylon-sheathed legs adorned the walls of the sleek Art Deco building. Du Pont's live exhibit further featured a brace of beautiful models playing an endless, girlish game of tug-of-war with an infinitely elastic nylon stocking.

As *Collier's* breathlessly reported, "Those Du Pont hostesses who pranced around the Du Pont exhibit all day" were more than mere window dressing: They were market testers in disguise. "They rinsed their [hose] out at night, and reported weekly on them in detail to headquarters in Wilmington."

Du Pont was equally proud of a slew of other new plastic products, including Lucite, the new crystal-clear polymethyl methacrylate plastic, introduced in the same year (1937) as Rohm & Haas's rival Plexiglas. The Wonder World of Chemistry displayed a glittering, giant nine-and-a-half-inch "diamond" cut from a block of "sparkling Lucite," while a "modern lamp of crystal-clear Lucite, intended merely as an incidental piece of furniture, received many an admiring ooh and ah from visitors," according to one contemporaneous report.

A guaranteed crowd pleaser was the Butacite exhibit, in which every ten minutes an enormous solid steel ball suspended twenty feet above a sheet of glass, hung horizontally, came crashing down on the glass with a resounding wham! But the new "laminated safety glass" didn't shatter because its Butacite interlayer, made of the new polyvinyl butyryl, kept the glass "integral" and prevented it from splitting into lethal shards.

As a synthetic understudy to Miss Chemistry, the marketing whizzes in Wilmington created an ancillary fantasy figure, the Test-Tube Lady, also known as Princess Plastic. The text accompanying her exhibit read: "From the tip of her heels to the topmost curve of her new hat . . . and fashionably speaking only, mind you . . . the modern woman is more and more a product of the plastic laboratory. . . . "

Princess Plastic sported "bright, sleek heels covered in Pyraheel, a Du Pont plastic heel covering, which will match almost any leather effect and still resist scuffs." Her hat, delicately trimmed with "Du Pont Plastacele, in glistening black," belied the whole image of bright colors "which come to life in this Du Pont plastic."

"Plastic, too, are her slide fasteners, shoe bows, lipstick container, and even her crystal-clear walking stick, made entirely of new Du Pont Lucite. And notice, especially, her necklace and bracelet, with entirely new pearl effects made possible with Lucite."

All around the fair, plastic played an increasingly critical role in realizing the fabulous "World of Tomorrow." At Donald Deskey's Communications Exhibit, an elaborate display called "Man the Communicator" illuminated the role played by communications in modern society. Depicting electronic communications as the central nervous system in a vast "body electric," "man" was symbolized by a twenty-foot-tall plastic head, speaking through the symbols of seven instruments of communications, which materialized on a plastic disk in front of him before being projected onto a thirty-foot-high plastic globe suspended at the opposite end of the exhibit hall.

One of those seven electronic muses was television. Along with nylon, TV made its vaunted debut at the fair. At the RCA building, long lines waited patiently to catch a glimpse of a transparent television set on display, whose crystal-clear Plexiglas skin revealed the mysterious workings of the cathode-ray-tube transmitter within.

If cellophane had turned "a yen for transparency into a national obsession," Lucite and Plexiglas turned that yen into a jones. The Philadelphia-based Rohm & Haas Company, developer of Plexiglas, produced an elaborate animated display in the Hall of Industrial Science demonstrating a unique property of its new plastic: a remarkable capacity to transmit light around bends and curves without diffusing it to the sides. It could transmit and intensify a beam of light by focusing it to a pencil point, or "pipe" it around curves, making it a favorite of nightclub designers. The company even sponsored a competition among modern sculptors to develop artistic applications for Plexiglas. First prize was awarded to the young Stevens Institute of Technology graduate Alexander Calder, third in a generation of sculptors from the famous Philadelphia artistic family.

Rohm & Haas displayed a Plexiglas flute, "said to have splendid formal qualities and is entirely suitable for concert purposes." The same could not be said about the Plexiglas violin, which looked ethereal, and sounded like something not quite of this planet. The Home Center featured a full-sized Plexiglas house, while at the GM Futurama, visitors climbed out of miniature moving cars to gaze in astonishment at a full-size, fully featured Plexiglas Pontiac (body by Fisher), a Torpedo 6 with white sidewall tires rotating on a transparent plastic pod in the sleek, streamlined lobby of the Highways and Horizons Exhibit.

"This car comes complete with windows which can be raised and lowered," proclaimed a Plexiglas plaque at its base, "as well as doors that can be opened and closed. It might be driven out on the highway should occasion demand."

Sheer Magic

In one of the most effective marketing teasers in consumer history, Du Pont forced the women of America to wait a year or more before they could actually lay their hands—or legs—on an actual pair of nylon stockings. That "yen for transparency" had likewise taken over the full-fashioned hosiery market. The first trial sale took place, of course, in Wilmington, Delaware, a year to the day after Stine's announcement. In preparation, full-page ads appeared in the Wilmington *News-Journal*.

Women of Wilmington Are Invited to Inspect and Purchase the Stockings Everyone Has Been Talking About.

Only three pairs could be sold to any one customer, who were required to supply a sales clerk with a valid local address. Determined to get their hands on a pair of nylon stockings by hook or crook, thousands of wealthy women from the world beyond Wilmington frantically booked hotel rooms inside the city limits so they could supply the requisite local address. Those not quite so obsessed tied up phone and telegraph wires attempting to place phone and wire orders from all over the country, but even those able to get through were politely refused. The four thousand pair of stockings on sale had been knitted by a tiny cabal of hosiery knitters from a limited supply of raw fiber, representing the entire production run of the pint-sized pilot plant pumping around the clock at the Experimental Station.

Priced at $1.15, $1.25, and $1.35 (depending on gauge) the first batch of nylon stockings were much more expensive than comparable silk hose, which retailed at 79 cents a pair. No customer was allowed to walk out of any of the six participating stores with stockings in hand—all purchases had to be delivered, so the company could keep close tabs on buyer reaction. Even saddled with all these restrictions, the entire stock of four thousand pairs was sold out in three hours.

Every Wednesday for the next six months, some twenty lingerie shops and department stores in Wilmington would quietly put new stocks of nylon stockings on sale. As *Time* condescendingly reported, "As soon as the doors open the women of Wilmington rush in like so many hens at feed time." Though no customer was allowed to purchase more than three pairs at a time, "Nightfall finds few pairs of nylon hose left on store shelves anywhere in Wilmington." In search of an explanation for this remarkable phenomenon, *Time* begrudged the notion of a miracle material. "One reason Wilmington women like them is because only Wilmington women can buy them."

The first national test would come on "N-day"—the morning of May 15, 1940, when fashionable women across America, determined to be first in their set to show off a pair of nylon stockings, awoke before dawn to take their place on lines snaking around selected downtown department stores for blocks. As in Wilmington, only a limited allotment of stockings had been provided to an exclusive group of retailers—typically only one per major market. The scene duplicated at hosiery counters across America was invariably described as "pandemonium"—though "stampede" came in a close second. Even with sales strictly limited to one pair per person, the entire national quota of five million pairs was swept out of stock before sundown.

An awestruck *Fortune,* predicting that if nylon mania could be sustained, nylon might quickly surpass rayon's $250 million-a-year business, got down to brass tacks by conducting an informal, independent assessment of nylon's relative strengths and weaknesses. In a probing report from the synthetic front, *Fortune* concluded that though nylon stockings might superficially resemble silk, they "have a personality all of their own." From a strictly tactile point of view, nylon was "colder, harder, and smoother to the touch than silk." Some women even described nylon stockings as "clammy." They washed and dried in about half the time of silk, but their nonabsorbent character caused some wearers to complain that they become slippery in the rain, or with perspiration. As for being "strong as steel," nylon threads seemed to snag about as easily as silk, but the thread didn't break quite so easily. Once broken, nylon threads ran faster than silk.

One factor in nylon's stunning success that could not be overlooked fell under the rubric of "sheer magic": its essentially sexy, slinky, sleazy character. Silk seemed virtuous; nylon naughty. Nylon mania kept the gutter-level male minds of America focused on one thing: female gams. "On the basis of scattered reactions in the early days of nylon's national sale," *Fortune* forthrightly concluded, "American womanhood seems to be already divided into a small, violently anti-nylon camp on one side and a larger, enthusiastically pro-nylon group on the other." Nylon, in short, had legs.

For the first time in history, a synthetic material had been judged by the public more valuable than its natural counterpart. If there remained any doubt about nylon's appeal, it was dispelled by the extraordinary number of cheap hucksters across the country caught trying to pawn off silk stockings *as* nylon.

The *New York Times* hailed nylon's stunning success as an indication of the increasingly profound role played by synthetic materials in modern life.

TRIUMPH OF SYNTHESIS

If nylon were merely another rayon, the advertisements that announce its commercial introduction in the form of stockings might call for no special comment. But nylon happens to mark a new stage in the development of synthetic chemistry. As such it is a triumph of the industrial laboratory and a new approach to the problem of changing the environment of men.

Usually a synthetic is a reproduction of something found in nature. A dye, a perfume is pulled apart and its construction discovered, whereupon the chemist proceeds to build it up in his own way. This nylon is different. It has no chemical counterpart in nature. Structurally, it bears no more resemblance to silk or rayon than cotton does to hair. To say it's made of coal, air and water explains little. The air is actually nitrogen, the essential ingredient of protein—beefsteak, white of egg, milk. Technically, nylon is a polyamide, which means that it is the first step in the production of synthetic proteins or food— momentous for that reason alone.

The synthetics are changing life imperceptibly—but as profoundly as it was ever changed by the machine. The gaudy fountain pens and cigarette holders, the lacquers, the well-designed and attractive costume jewelry in the five-and-ten stores, the curtains at the living room windows, the steering wheels that we clutch in our cars, the drugs that cure us from some of the more deadly diseases, the dyes that outdo the rainbow, the camphor that once came from trees, the musk that once cost $700 a pound, the "glass" in the bathroom—all are synthetics.

To the man in the street, "synthetic" still means some tricky substitute for something authentic—an unfortunate relic of prohibition when raw and fiery alcohol was inexpertly flavored to produce a potent but unconvincing gin. It is time that we give synthesis its true meaning—a putting together, a control over matter so perfect that men are no longer utterly dependent upon animals, plants and the crust of the earth for food, raiment and structural material.

Purity Hall

December 18, 1926
To: Executive Committee

From: Dr. C. M. A. Stine, Chemical Department Director

We are including in the central Chemical Department's budget for 1927 an item of $20,000 to cover what may be called, for want of a better name, pure science or fundamental research work. The purpose for which this sum is requested represents a sufficiently radical departure from previous policy that it seems advisable to present the matter in this special letter. . . .

In submitting to Du Pont's executive committee a proposal for a budget to embark on a program of "fundamental research," Dr. Charles M. Stine was putting himself way out on a limb. To build support for such a "radical departure from previous policy," Stine sought to appeal first and foremost to his superiors' competitive instincts. "Not only is [pure science] fostered to a considerable extent by foreign industries, particularly in Germany," Stine sagely advised, "but also by certain concerns in this country, notably the General Electric Company."

Germany's unconditional surrender to the Allies during World War I, called by many "the chemists' war" had been inevitable after a successful naval blockade by the Allies produced desperate shortages of key raw materials. Lacking the colonial support arch rivals England and France had long relied on for critical imports of raw materials, Germany had been forced to find another way out: the passionate pursuit of chemical synthesis.

With the German dyestuffs and pharmaceuticals industry already far in advance of its rivals on the Continent in the synthetic production of vital nitrate-based fertilizers and explosives, World War I had witnessed a massive effort to foist *ersatz* (synthetic) products on the German public as a means of conserving scarce raw materials for its fighting forces.

After the armistice, even as Germany pledged itself to permanent peace, the formation in 1925 of the vast I.G. Farben chemical conglomerate (Interessengemeinschaft Farbenwerke) could be seen as part of a concerted strategy to maintain German synthetic strength in the event of a future war.

During the late 1920s, and throughout the 1930s, even as the German economy collapsed, I.G. Farben liberally underwrote basic or fundamental research in chemical synthesis as the most desirable way of building up a strategic reserve without violating Allied sanctions. Starting in 1927 (the same year that Du Pont ultimately funded its own research program under Stine) Farben assembled a crack team of organic chemists to exclusively pursue promising lines of research in polymer chemistry.

Over the following decade, the Farben team refined modern polymer structure theory (the rudiments of which had been laid down in Germany by Hermann Staudinger) into a series of practical methods of constructing specific molecules according to detailed chemical blueprints. Farben chemists succeeded in synthesizing a new polymer on the average of every day for ten years.

In Britain, the globe-spanning juggernaut Imperial Chemical Industries (I.C.I.) was hastily cobbled together in belated response to the looming threat posed by Farben. In the new global age of chemical cartels and combines, the one American organization capable of competing with Farben and I.C.I. was clearly Du Pont. Which made it all the more critical, in Stine's judgment, for Du Pont to join Farben and I.C.I. in underwriting fundamental research.

A twenty-year veteran of Du Pont, Stine was well versed in the strategic imperatives of the chemical industry. He had helped to develop dyes and explosives for military use, and personally manufactured TNT for the navy during World War I. After rising through the research ranks to be appointed chemical director in 1924, Stine considered it his duty not just to his company but also to his country to persuade his superiors to heed President Herbert Hoover's recent raising of the red flag over the sorry state of fundamental research in the United States.

To further bolster his argument, Stine cleverly employed a

straightforward analogy any Du Pont official could intuitively grasp: that if Du Pont failed to act now, there might soon come a time when the supply of "scientific facts" used by industrial scientists to "solve practical problems" would run dry in America. Even worse, if domestic shortages of intellectual raw materials were to occur in the United States, the most likely beneficiary of Yankee shortsightedness would be Germany.

Stine's soberly persuasive memo achieved its desired results, and more. In April 1927, Du Pont's executive committee allotted Stine's chemical department $25,000 a *month* (a substantial raise over Stine's request for $20,000 a *year*) to underwrite basic research in organic chemistry. As an added bonus, they granted Stine an immediate one-time payment of $125,000 to finance the construction of a new "pure science" laboratory at the Experimental Station in Wilmington.

Well before it was completed, the new building was dubbed "Purity Hall"—a nickname grudgingly bestowed, one of its occupants later insisted, "in a spirit of cheerful derision, tinged with a trace of envy." Stine lost no time in actively canvassing the country in search of able-bodied recruits to fill the new building. His stated goal: "To hire men of proven ability and recognized standing in their respective fields."

After conducting a preliminary round of interviews, however, and putting out a few tentative feelers, Stine soon found himself wielding a financial battering ram against the outer ramparts of the ivory tower. After being flatly turned down by Professor Roger Adams of the University of Illinois, one of the leading organic chemists of his day, followed by Professor Henry Gilman of Ohio State, Stine was forced to conclude that no scientist of worldwide renown or academic standing would be likely to accept even his most lucrative offer, out of fear of risking his academic reputation. It wasn't that Du Pont was perceived as sordid or sinister but that the pursuit of scientific truth and the pursuit of profit were regarded—often correctly—as antithetical goals.

Stine was forced to substantially lower his sights and follow the humbler recruiting strategy adopted by General Electric, which targeted "men of exceptional scientific promise but no established reputation" for staffing even its most senior research positions. One advantage of this more low-key approach, Stine duly conceded, was that "all lines of research could be largely determined by us."

After cordially declining Stine's offer, Professor Adams of Illinois was gracious enough to recommend a young protégé of his for the job: a thirty-one-year-old organic chemist and instructor at Harvard, Wallace Hume Carothers. Carothers's qualifications could easily be summarized; they had been succinctly set forth by the Illinois faculty in 1924, when on the occasion of awarding him his Ph.D. in organic chemistry, they had officially declared Carothers "one of the most brilliant students ever to be awarded a Doctor's Degree in Chemistry by this faculty." Du Pont, Adams insisted, would be lucky to have him.

Doc II

Wallace Carothers's elementary school classmates in Des Moines had been so impressed by his ability to wire electric doorbells and to wind coils for crystal-set radios inside old Quaker Oats boxes that they called him "Doc" for short. When he was five, Wallace Carothers had moved with his family from his mother's farm in Burlington, Iowa, to the bustling metropolis of Des Moines so that his Scottish-born father Ira, a former country school teacher, could take up a teaching job at the Commercial College, a small private business school.

The oldest of four children, Wallace acquired a lifelong love of music from his mother Mary, who also instilled in her children an abiding love of literature. This led Wallace to devour, before turning ten, the collected works of Mark Twain, *Gulliver's Travels*, and one of his favorite books: a popular life of Thomas Edison. If he had his life to live over again, Carothers claimed, he would have liked to have been a concert pianist. But lacking any notable musical ability or talent, he had stubbornly refused to learn a musical instrument on the grounds that it would have been a waste of time unless he could excel at the highest level. He also had his masculine image to consider. "Learning to play the piano," he would later confide to a close friend, "just wasn't the sort of thing a regular fellow should spend his time on."

After graduating in 1914 at the top of his class at North High School in Des Moines, this "regular fellow" enrolled at his father's Commercial College, where he easily raced through a two-year course in business methods and accounting in a year. Lacking the means to support further studies without taking a job, he used his business degree to land a teaching post in the commercial department of tiny Tarkio College in Tarkio, Missouri, where he also matriculated as an undergraduate majoring in chemistry.

While teaching full-time, Carothers easily placed at the top of his class in every course, from organic chemistry to English literature. His writing and literary skills were so outstanding that in his sophomore year he was awarded a teaching assistantship in English, which allowed him to drop business teaching. He completed all his required courses in chemistry by the end of his sophomore year, allowing him to temporarily replace the departing Dr. Arthur Pardee, head of Tarkio's chemistry department, until a full-time replacement could be recruited. Though only a junior, his skills as a chemistry instructor were amply confirmed when all four of the chemistry majors he taught in his senior honors class went on to complete doctoral work in the field.

After being granted an exemption from military service in World War I due to a goiter condition, Carothers graduated with a B.S. from Tarkio in 1920. After earning his M.A. from the University of Illinois under the illustrious Roger Adams, he once again ran out of money. Dr. Pardee, his old professor from Tarkio, came to the rescue by offering him a one-year teaching stint at South Dakota, providing him with sufficient funds to reenter the University of Illinois the following fall.

Carothers's chronic financial problems were temporarily solved when Adams secured for him the prestigious Carr Fellowship, the highest academic award offered by the University of Illinois to budding chemists. Financial independence not only permitted him to complete his doctoral studies without worrying about money but also gave him enough surplus to indulge in his two other great extracurricular passions: late-night billiards and strong coffee.

Roger Adams on Carothers at Illinois:

> Carothers was deeply emotional, generous, and modest, with a lovable personality. Although generally silent in a group of people, he was a brilliant conversationalist when with a single individual, and quickly displayed his broad education, his wide fund of information on all problems of current life, and his critical analysis of politics, labor problems and business, as well as music, art and philosophy.

But there was a darker side to Carothers, one well known to Adams, which revealed itself even in his undergraduate days in "frequent bouts of depression, which grew more pronounced as he grew older." Those dark spells frequently began with a sensation of profound boredom and listlessness, when even his groundbreaking research (on the action of platinum oxides as catalysts for reducing organic compounds) struck him as nothing more than make-work. "Graduate school," Carothers confided in a letter to a friend, "contains all the elements of adventure and enterprise which a nut screwer in a Ford factory must feel on setting out for work in the morning."

He screwed in enough nuts to earn his doctorate in 1924, and after two years of teaching at Illinois was recruited by Dr. James B. Conant to join Harvard's newly invigorated chemistry department. When, years later, one of Harvard's overseers expressed mild misgivings about appointing Dr. Conant, a chemist, to be president of Harvard, a fellow overseer casually remarked that the renowned Harvard president Charles W. Eliot had also been a chemist.

"Yes," the first overseer gloomily replied. "But not a *good one*."

Conant was a good one. And he knew a good one when he saw one. Conant on Carothers at Harvard:

> In the brief space of time during which he was a member of the Chemistry Department, Carothers presented Elementary Organic Chemistry to a large class with distinction. Although always loath to speak in public even at scientific meetings, his diffidence seemed to disappear in the classroom. His lectures were well-ordered, interesting, and enthusiastically received by a body of students only a few of whom planned to make chemistry a career.
>
> In his research, he showed even at that time the high degree of originality which marked his later work. He was never content to follow the beaten track or to accept the usual interpretations of organic reactions. His first thinking about polymerization and the structure of substances of high molecular weight began while he was at Harvard.

Conant generously described Carothers's resignation from Harvard and "his acceptance of an important position in the research department of the Du Pont company" as "Harvard's loss, but Du Pont's gain."

After being approached by Stine to consider the position of head of research for Du Pont's new pure science division, Carothers was, not surprisingly, ambivalent. The essential difference between Carothers and the scientific eminences interviewed for the job before him was that he was at the outset, as opposed to the peak, of a stellar career; and that years of financial hardship had provided him with an acute appreciation of the value of a dollar. Fortunately, Stine had another ace up his sleeve of far greater value than money. He could guarantee that the Du Pont position would come blissfully free of teaching duties. Though Carothers had once described Harvard as "an academic paradise," in real life he did not like to teach.

As an opening move in the chess game, Carothers demanded explicit guarantees that he could continue to work on "the thermal decomposition of ethylmetal compounds" free of corporate interference. "The problem," he freely conceded, "has some explicit bearing on theoretical chemistry, but none so far as I know [will] be of any practical use." By deliberately deploying phrases like "of no practical use," Carothers was clearly testing the limits of Du Pont's stated commitment to "pure science." Just how "pure" was Du Pont willing to be?

It was the right question, and in attempting to answer it, Stine proved unwilling to promise Carothers the moon. While Carothers would be free to pursue any lines of inquiry he pleased, his future at Du Pont would be "in no small degree dependent on [his] capacity for initiating and directing work that we consider worthwhile."

In a rambling reply to Stine, Carothers outlined his remaining reservations. Since his primary goal was personal scientific advancement, he had no choice but to weigh Du Pont's offer against the advantages and prestige of Harvard. In Harvard's court, Conant had recently succeeded in securing enough funds to construct a new chemistry laboratory at Harvard. If his current research panned out, Carothers stood a fair chance of getting his teaching load drastically reduced. He further remained seriously concerned about Du Pont's capacity to "suppress the development of an investigation"— if it was determined, for example, to have no positive impact, or even a negative impact, on the bottom line.

The one final caveat he chose to raise was a personal as

opposed to a professional matter. He was frankly concerned about his ability to adapt to Du Pont's straitlaced corporate culture, given that he openly suffered "from neurotic spells of diminished capacity which might constitute a more serious handicap there than here." To Stine's lasting credit, this candid admission seems to have had no substantial effect on the progress of negotiations.

In a last-ditch effort to yank Carothers off the fence, Stine dispatched one of his ablest assistants, Hamilton Bradshaw, to Cambridge for a talk with his target. Bradshaw and Carothers hit it off at first sight. The mere fact that a kindred spirit like Bradshaw had not only survived but thrived at Du Pont came close to eliminating Carothers's last lingering doubts. But in the event that Carothers still could not commit, Bradshaw conveyed a final message from Stine: Du Pont would be willing to up its ante by 20 percent, to an annual salary of $6,000, if that would help him to arrive at a final decision. The combination of Bradshaw and a sweetened pot carried the day for Du Pont. Ten days later, Carothers accepted Stine's offer.

After finishing up the semester at Harvard, he moved down to Delaware at the end of the year (1927) to settle into his spanking-new laboratory in the spanking-new Purity Hall. Du Pont's Experimental Station was a far cry from a stately Harvard quadrangle. It has been described by one observer as "a group of small buildings having an inelegance akin to shanties, clustered about larger brick buildings which by comparison would have been called imposing." Carothers's laboratory was officially denominated Building 228. "The magnitude of this number was a source of pride to the chemists," this observer maintained, "because it indicated that their company was large enough to have a building numbered that high."

Carothers's staff, all young and mainly bachelors, were fiercely dedicated to their work. But they fancied themselves as daring young heroes, living on the edge of mortal danger. Because the staff was dealing with unknown chemical reactions, Purity Hall (like all laboratories of the presprinkler era) came equipped with escape chutes leading from upper-story windows down to the ground. Staid members of the Experimental Station were often taken aback to see Carothers's scientific shock troops exiting their labs by sliding out windows on their hip pockets, in a flamboyant display of *esprit de corps* worthy of aces in a crack air squadron. Shortly after settling in at Du Pont, Carothers wrote to his old colleague Jack Johnson, at Cornell:

Feb 14, 1928

Dear Jack,

A week of industrial slavery has already elapsed without breaking my proud spirit. Already I am so accustomed to the shackles that I scarcely notice them. Like the child laborers in the spinning factories and the coal mines, I arise before dawn and prepare myself a meager breakfast. Then off to the terrific grind arriving at 8 just as the birds are beginning to wake up. Harvard was never like this. From then on I occupy myself by thinking, smoking, reading and talking until five o'clock.

For all his ironic Dickensian references to child laborers and shackles, Carothers was evidently impressed by the physical plant, which he described as "rather spiffy to one used to Boylston Hall." He also reveled in the generous level of material support, combined with an apparent corporate indifference to mundane details like work schedules, spending limits, and even research goals.

As for funds, the sky is the limit. I can spend as much as I please. Nobody asks any questions as to how I am spending my time or what my plans are for the future. Apparently it is all up to me. So even though it was somewhat of a wrench to leave Harvard . . . the new job looks just as good from this side as it did from the other. . . .

Carothers felt free to describe in detail his plans to Johnson while keeping his superiors in Wilmington largely in the dark. He was plainly inspired by the prospect of using the superb facilities, staff, and tools at his disposal to prove some of the more controversial theories of Hermann Staudinger, who in 1920 had outlined the basic theory of polymer chemistry by claiming that "polymeric molecules"—today we would call them polymers—are composed of molecular chains of practically limitless length and molecular weight, held together by conventional covalent bonds.

This view was in total opposition to the classical theory set forth by Scottish chemist Thomas Graham in 1861, whose Graham's law held that any substance whose molecules could pass through a very fine membrane—like salt or sugar—were "crystalline," while substances like rubber or gelatin, incapable of passing through the same membrane, believed to be constructed of larger molecules, were classified as "colloids." Though Graham was wrong about this distinction, as the father of colloidal chemistry he did discover dialysis.

"Colloids" were vaguely conceived as collections of smaller molecules held together by some mysterious "electrical" force. Though Staudinger would be awarded the Nobel Prize in chemistry in 1953 for his polymer theory, in the early days of his revolution his radical views were met with contempt and derision from orthodox chemists.

"My dear chap," the renowned German organic chemist Heinrich Wieland once told young Staudinger, "give up your ideas on big molecules. There are no organic molecules with a molecular weight of more than 5,000. Just you clean up your products and they will crystallize and reveal themselves as low-weight molecular compounds."

But Staudinger refused on principle to "clean up his products," and instead chose to challenge the system by conducting a series of groundbreaking experiments that compellingly convinced even his most committed skeptics that the forces holding together a compound like rubber are no different from the covalent chemical bonds holding together lower-molecular-weight compounds.

A committed proponent of Staudinger's theories from the outset, Carothers was personally convinced that the notion of some mystical unanalyzable "force" holding together high-molecular-weight compounds was "too strongly suggestive of the 'vital' hypothesis, which preceded the dawn of organic chemistry, to be seriously considered." By the "vital hypothesis," Carothers was referring to the belief that organic compounds were governed by a "divine force" qualitatively as well as quantitatively different from that controlling inorganic compounds.

One of Carothers's colleagues at Purity Hall, Dr. Julian Hill, would later credit Carothers with "laying to rest the ghost . . . that polymers were mysterious aggregates of small entities rather than true molecules."

To Jack Johnson, Carothers further outlined his experimental target:

One of the problems which I am going to start work on has to do with substances of high molecular weight. I want to attack this problem from the synthetic side . . . to synthesize compounds of high molecular weight and known constitution.

Bodies by Fischer

The highest molecular weight obtained by synthesis up to that time for a polymeric molecule was 4,200, achieved by the German organic chemist Emil Fischer. By comparison, the molecular weight of common table sugar is only 342. Hemoglobin possesses a molecular weight of 6,800, and rubber, one of the largest natural polymers, weighs in at an astonishing 1,000,000.

As Carothers eagerly wrote Johnson: "It would seem possible to beat Fischer's record of 4,200. It would be a satisfaction to do this, and the facilities will soon be available here for studying such substances with the newest and most powerful tools."

Staudinger's theories remained largely unconfirmed empirically because the methods of measuring molecular weight available to him—elevating the boiling point or depressing the freezing point of a compound—though they worked well enough for small molecules, had not worked for large polymers. Staudinger improved matters somewhat by devising a method of measuring the viscosity and density of polymeric solutions. But until Carothers stepped into the picture, no one had been able to synthesize molecules in the test tube any larger than those achieved by Fischer.

That acids react with alcohols to form esters is well-known to most first-year chemistry students. Carothers's radical first step in combining the two was to systematically build large-chain molecules—polyesters—by using "difunctional" molecules as building blocks, which come equipped not with one alcohol or acid group at each end, but two—at both ends. "Difunctional" or "polyfunctional" molecules are capable of reacting continuously with each other to set off chain reactions, which build up longer and longer chains of theoretically infinite molecular weight.

Not long after establishing himself at Purity Hall, Carothers was pleasantly surprised to find out that Du Pont had been constructing resinous polymers for high-resistance paints using a similar process for years—a happy coincidence that only confirmed his optimistic belief that there were additional benefits to be had from working in close proximity to the practical side of organic chemistry.

The big polymers Carothers synthesized in those early years would be later described by a Du Pont superior as "recurring structural units which might be crudely pictured as hooked together by a chain of paper clips." The paper clips were the conventional covalent bonds Staudinger and Carothers were convinced held together *all* molecules, even the high-weight variety. After a year of active synthesis, Carothers handily beat Fischer's record by creating a solid ester of 5,000. This feat not only earned Carothers and Du Pont some much-needed prestige but laid the groundwork for Carothers's subsequent confirmation of Staudinger's theoretical framework in the lab at Wilmington.

In 1929, Carothers published a classic paper describing his long-chain polyesters in the *Journal of the American Chemical Society.* Later that year, at the precocious age of thirty-three, he was appointed associate editor of the same journal. The following year, he was appointed editor of the journal *Organic Synthesis.* For an "industrial chemist" to hold high-level positions in "pure" chemistry was virtually unheard of until he broke the age-old barrier between academia and industry. In 1936, when he was elected to the National Academy of Science, he was the first industrial chemist ever to be so honored.

By the end of the twenties, as Stine proudly gave himself a pat on the back for launching a program "gaining an increasing recognition in the scientific world," Carothers was operating at full intellectual throttle, ably backed by a handpicked staff of eight full-time research assistants, all recruited straight out of graduate school. This elite group had become so accustomed to working in an atmosphere of unfettered freedom and dashing camaraderie that when the stock market crashed in October 1929, they were utterly unprepared to face the consequences.

The first sign that the slump was exerting any negative effect on Du Pont's commitment to pure research was that the company began to exert unsubtle pressure on Carothers and his carefree young staff to produce something more tangible than "pure ideas." Fortunately for the embattled denizens of Purity Hall, April 1930 was to be a month aptly described by industrial historians David Hounshell and John Kenley Smith Jr. as a *"mensis mirabilis"* (miraculous month) in the history of industrial research.

In that one month, Carothers's group would uncover the first synthetic rubber and lay the experimental groundwork for nylon. The driving force behind this remarkable metamorphosis from a relaxed Platonic symposium into a lean, mean production machine was Dr. Elmer K. Bolton, the newly appointed assistant director of Du Pont's chemical department. A recent arrival from dyestuffs, where he had directed research on synthetic rubber, Bolton was no believer in pure research but a firm proponent of aggressively pushing the scientists under his command to achieve bankable results "in the shortest time with the minimum expenditure of money."

But the practical-minded Bolton found himself forced to defend to the hilt even his nuts-and-bolts research program in the new cost-conscious climate. Willis Harrington, Bolton's boss at the dyestuffs department, would later admit to deliberately tweaking the hard-headed Bolton by repeatedly greeting him with the question: "What have you to show lately by way of accomplishment to justify your existence?"

Fortunately for Bolton, he would in short order have a lot to show for his efforts. For over a decade, Du Pont dyestuffs researchers under Bolton had tried and failed to synthesize a molecule that could compete with rubber in terms of elasticity. In early 1930, within weeks of taking full responsibility for activities at Purity Hall, Bolton asked Carothers to take a closer look at the polymer DVA (divinylacetylene), an extremely short chain composed of three acetylene molecules, which had been considered the most promising of all the candidates.

Under Carothers's direction, Arnold Collins obtained a transparent, yellowish DVA film, from which Carothers decided to isolate the impurities in the hope that they might reveal something about the nature of the compound itself.

After distilling an even cruder, yellower DVA film, Collins recovered a liquid (later called chloroprene) that spontaneously polymerized into a new, hitherto unknown compound.

On April 17, 1930, Collins excitedly wrote in his lab notebook that he had managed to solidify an emulsion of chloroprene "into a white, somewhat rubber-like mass, [which] sprang back to its original shape when deformed, but tore easily."

This "white rubber-like mass" was ultimately marketed by Du Pont under the brand name Neoprene. Once Neoprene's practical application was handed off to a more pragmatically minded team for commercial development, Carothers and his staff went on to co-author twenty-three papers systematically describing the basic chemistry of chloroprene and related compounds, which in his typical self-deprecating way Carothers would later describe as "abundant in quantity but disappointing in quality."

The most important thing about Neoprene was that it saved, if only for the time being, a goose just beginning to lay a multitude of golden eggs. And if Neoprene wasn't enough to keep Purity Hall in the good graces of the executive committee, later that same month, Dr. Julian Hill, a recent recruit from MIT, set out to synthesize a polymer of molecular weight higher than 6,000.

Breaking the 6,000-molecular-weight barrier loomed as an apparently insurmountable obstacle to Carothers's goal of producing a true "superpolymer" in the lab. Convinced that the production of excess water during the polymerization process posed the major impediment to chain growth during the reaction, Carothers assigned Hill to the investigation of a variety of methods of eliminating water.

At a scientific conference held within months of receiving this key assignment, Hill attended a lecture in which a new piece of equipment called a molecular still—a high-vacuum distillation device demonstrably more efficient than any of the conventional stills used up to that time—was described in some detail. As Hill later put it, a molecular still is "a device in which the distance between the distilling and condensing surfaces is shorter than the path of the gaseous molecules at low pressures. Escaping molecules are therefore captured irreversibly."

Upon returning to Wilmington, Hill constructed and modified the molecular still he saw on the lectern to suit the requirements of synthesizing a polymer of maximum molecular weight. He placed one of the polyesters earlier synthesized by Carothers during his first round of experiments (trimethylene ester of hexadecamethylene dicarboxylic acid—also known as 3–16 Polyester) into the still, added a little propylene glycol (the primary ingredient in antifreeze), heated this mixture to precisely 200 degrees centigrade, and maintained that temperature for a period of twelve days.

Lo and behold, after opening the still, Hill observed a hard, white opaque residue that appeared to be permanently cemented to the inside of the container. To get a better feel for this polymer, he heated a glass stirring rod and inserted its hot end into the mass, which obligingly melted around the rod. As the mass gradually cooled, it "froze" the rod into place, preventing it from being easily withdrawn.

Hill then heated the glass beaker until a thin plastic film of molten polymer formed on its inner surface, and was then able to pull the molten mass free from the glass. As he attempted to take out the glass rod, he gently pulled from the center of the viscous melt a cluster of long, thin streamers, which stretched from the end of the glass rod back into the beaker, like gossamer strands of hot taffy.

As the silken streamers cooled, they hardened into opaque, brittle filaments. Hill would later describe them as "easily broken." But if stress were gradually and gently applied to these solidified filaments, a thin opaque section of thread would appear, separating the two transparent sections. As Hill continued to gently tug at those strands, he found that he could easily stretch them to several times their original length. As the strands stretched, the transparent section between the opaque sections of thread gradually grew larger, until—with the fiber stretched to full length—both opaque sections would in time turn transparent. The most remarkable feature of these fibers was that after being stretched to about seven times their original length they became highly elastic. You could stretch the strands further, but once the tension was released, they snapped back like chewing gum.

As Hill later described this moment of sheer exultation:

> When all the unoriented filament is exhausted, and all of a sudden any further pulling threatens to cut your fingers, you have the sensation of feeling the molecules lock into place as they fall into parallel array, and the hydrogen bonds begin to take hold. . . .

X-ray diffraction analysis, and closer examination under polarized light, revealed a few curious events taking place at the so-called drawing boundary. As the strands stretched into long, highly elastic polymer chains, the molecules within them, which had been lying helter-skelter in the filament, were "reoriented" into compact bundles of molecules running parallel with the fiber axis. These so-called oriented polyester fibers were tough and pliable, and exhibited a remarkable tensile strength—comparable to that of silk. Elastic recovery—the ability to "snap back"—was superior to that of both silk and rayon. Not only did Hill's remarkable new fibrous compound compare favorably with silk in terms of both strength and elasticity, the molecular structure of the fiber, X-ray diffraction analysis revealed, was uncannily like that of silk. As an added bonus, Hill's "superpolymer" was soon found to possess a molecular weight of over 12,000, more than twice the weight of any polymer ever synthesized in a lab.

As Julian Hill later jubilantly recalled, "These results were not only of the greatest scientific interest, but they also gave birth to an enormous industry dedicated to the production of synthetic fibers." Rayon and cellulose acetate were not true synthetics but cellulose-based semisynthetics. As celluloid was to Bakelite, rayon and acetate were to this new fiber.

More than a few kinks would have to be worked out before Hill's taffylike polyester would be able compete effectively with its natural rivals on the open market. Compared to silk, 3–16 Polyester melted easily in hot water, dissolved in dry-cleaning fluid and alcohol, and became easily waterlogged. On the other hand, it displayed a remarkable elasticity, had extremely high tensile strength, and retained nearly all of that strength when wet.

Still, Hill's 3–16 Polyester might well have remained nothing more than a lab curiosity if Charles Stine had not been elevated, two months later, to Du Pont's executive committee, paving the way for Bolton to succeed Stine in the post of chemical director.

One of Bolton's first decisions as chemical director was to second the decision of his hard-driving lieutenant Ernest Benger that Carothers and Hill would not be permitted to publish their groundbreaking fiber findings until they had been "protected by a well-planned patent program." As opposed to the "publish-or-perish" imperative of the academy, Benger had instilled a new ethos: "publish *and* perish."

This prohibition against publication was in blatant violation of the pledge Stine made to Carothers three years before. Horrified by the mercenary motives of the new penny-pinching regime, Carothers forced Benger to consult their original correspondence.

Benger was appalled to find that Stine, as Carothers claimed, had explicitly granted Carothers the privilege of publishing theoretical findings without having to seek the approval of senior management.

As a practical expedient, however, Carothers knew on whose side his bread was buttered. Electing not to stand on principle, he agreed not to publish their momentous findings until after the relevant patents had been filed and approved. They were forced to wait a full year later, until September 1, 1931, to present their results to a thousand members of the American Chemical Society, who were holding their annual convention at the Statler Hotel in Buffalo, New York.

Abandoning all pretense at conventional scientific restraint, Carothers and Hill studded their statement before the 82nd Chemical Congress with flamboyant terms like "remarkable" and "spectacular" in describing the unique phenomenon they called "cold-drawing."

"Polymers of a very high molecular weight have been synthesized in the past," they conceded. "But the capacity to yield permanently oriented fibers of any considerable strength has not been observed." Since "spinning of continuous fibers is not possible until molecular weights of about 7,000 are reached," they pointed out, "the property of cold-drawing does not appear until the molecular weight reaches about 9,000 . . . while useful strength and pliability requires a molecular weight of 12,000."

In conclusion, Carothers and Hill were less than shy about claiming that "a clump of filaments rolled into a small ball and compressed showed a remarkable springiness resembling wool. In their elastic properties, these fibers are very much superior to any known artificial silk. . . . The experiments recently conducted clearly demonstrated for the first time the possibility of obtaining useful fibers from a strictly synthetic material." Yes, this was the dawn of a new era in science: one in which intellectual advancement would be cold-drawn by market forces.

The synthetic breakthrough made the front page of the *New York Times*.

CHEMISTS PRODUCE SYNTHETIC 'SILK'

DU PONT EXPERTS TELL CONVENTION AT BUFFALO
OF RESULT OF LONG EXPERIMENT

HIGH LUSTRE IS FEATURE

STRENGTH, PLIABILITY AND ELASTICITY OF NEW FABRIC ARE DESCRIBED AS REMARKABLE

Hitherto the various artificial silks, such as rayon, could not be made out of "strictly synthetic material." Rayon, for example, is pure cellulose . . . turned into rayon by substituting a machine for a silk worm. The new process makes it possible for the first time to begin with simple organic substances and build up from them to the more complex ones with molecular weights from up to four times greater than had been possible hitherto to create artificially.

One remarkable feature of this report was the indiscriminate use of two "hithertos" in one bluntly truncated lead paragraph. The *Times* went straight for the classic "castor oil and antifreeze" formulation, soon picked up across the country. "The Du Pont chemists began with a compound of ethylene glycol, an anti-freeze mixture, and diabasic acid which may be obtained by mixing castor oil under heat with an alkali base. . . . "
Time, too, went for the catchy antifreeze-and-castor-oil angle.

CASTOR OIL SILK
By heating castor oil and an alkali and mixing the result with the motor anti-freeze compound ethylene glycol, Wallace Hume Carothers and Julian W. Hill, Du Pont chemists, produced an artificial silk fiber. . . . The Carothers-Hill fiber is as lustrous as real silk, stronger and more elastic than rayon, as strong and elastic as real silk. It is too expensive to manufacture commercially, and is mainly a demonstration of chemical knowledge and skill. . . .

The Buffalo *Courier-Express,* adopting a more local perspective, preferred to emphasize the arresting fact, revealed for the first time at the chemists' convention, that "Fuel Can Be Made from Beer Waste."
Polyester 3–16 was mentioned only in passing, and in a derisory tone.

Lifeless matter is used to make this new product. . . . The result is a soft silk substance which some women think might be made into a dress. They might call it silk after it is treated by machines, but to the synthetic chemist it is hexadecamethylenedicarboxylic acid. Imagine going into a store and asking for a yard of that!

What would be of "undoubtedly greater interest to mothers and children," this intrepid science reporter insisted, "is that spinach may be taken in the form of a pill." For some unknown reason, the *Courier-Express* had it in for Du Pont.

Time, in its punchy effusion, had hit the nail right where it hurt. That this new hallmark synthetic fiber remained "too expensive to manufacture" and was "mainly a demonstration of chemical knowledge and skill" was, not surprisingly, nothing but a humiliation for practically minded Benger and Bolton. With the hot breath of senior management trained down their necks, Carothers and Hill returned to Wilmington from Buffalo newly motivated to begin the systematic exploration of a series of compounds chemically related to Polyester 3–16, in the hope of finding a compound that combined a capacity for "cold-drawing" with the elusive high melting point.

Polyamides (which combine an amine with an acid instead of an ester) have higher melting points than polyesters. But the polyamides Carothers and his team synthesized could not, for some inexplicable reason, be cold-drawn. Carothers became convinced— erroneously—that the same chemical attributes that permitted cold-drawing to take place also required low melting points.

"If there were some means of synthesizing and spinning [the polymer] at the same time, as perhaps a silkworm would," he speculated, "then it might be possible to get around this difficulty." But rather than spur him to further action, such moody musings only sent him further down the road to the pessimistic conclusion that the synthetic silk search should be abandoned—for now.

A position of passive resistance could be detected lurking behind this defeatist attitude. If Benger and Bolton were determined to force the mighty Carothers to pursue profit, he would show them that profit and purity could not be so easily reconciled. With barely a nod to Benger and Bolton, Carothers defiantly plunged into a subject of apparently only theoretical interest: the tendency of certain polyesters to form carbon rings instead of chains.

Ironically, the detour that Carothers chose to take at this rebellious juncture led to yet another unforeseen commercial success. Like some reluctant Midas, everything Carothers touched in the hope that it would yield a crop of pure ideas somehow managed to be worth its weight in gold. In the course of his investigation of these so-called cyclic compounds, Carothers struck pay dirt. He couldn't help noticing that this cyclic compound exuded a distinctly musklike aroma. Du Pont promptly produced it as "Astrotone"—the first synthetic musk.

By late 1933, having completed his classic research in polymer formation, Carothers found himself at a dead end—theoretically, spiritually, practically, emotionally. For the first time in a remarkably productive five-year career as an industrial chemist, Carothers had nothing to do.

In quiet despair over his future at Du Pont, Carothers sounded out James Conant—recently elevated to Harvard's presidency—about the possibility of returning to Cambridge to teach chemistry. But he had just bought a new house in Wilmington for his parents, recently bankrupted by the depression, and within a few weeks thought better of this fantasy escape route.

In early 1934, as Bolton cagily and fretfully observed Carothers's vacillation, he began prodding his reluctant genius to take a second crack at synthetic silk. Carothers himself later wrote, "In early 1934 I decided it [the polyamide project] was worth one more effort." Bolton later recalled Carothers stepping into his office to announce, "I think I've got some new ideas."

These new ideas consisted mainly of the realization that Hill's molecular still, for all its advantages, might have become the problem, not the solution. If Carothers could only produce a high-melting-point polyamide by using a purified amino acid instead of the acid itself, he might be able to dispense with the molecular still.

On March 23, 1934, Carothers asked one of his assistants, Donald Coffman, to prepare a fiber from a particularly promising ester. Coffman devoted five weeks to preparing the mixture, which readily polymerized into a compound that gave every sign of possessing a truly massive molecular weight. Coffman was forced to jump back as the new substance dramatically revealed itself as a formidable superpolymer: After forming a film on the inside of the flask, it suddenly smashed it to smithereens.

The following morning, Coffman drew a long, slender monofilament from four grams of the polymer and recorded in his lab notebook a potentially earth-shattering result: After he heated it in a bath to 200 degrees centigrade, its melting point appeared to be 195 degrees centigrade—precisely the level Carothers had stipulated as the experimental goal.

Now the acid test became whether the compound could also be cold-drawn. Just as Hill had done four years before, Coffman inserted a heated glass stirring rod into the molten mass, withdrew it, and patiently pulled out a cluster of long fibrous strands that struck him as "fairly tough, not at all brittle" and —most importantly—capable of being "cold-drawn to yield a lustrous filament."

When Carothers examined the compound, he later reflected that this would have been a good time to shout out "Eureka!"—if he had been that sort of chap. Instead, after drawing a few grams of Coffman's compound into a hypodermic needle, he strode calmly into Bolton's office, leaned over his master's desk, and calmly squirted a few drops of the solution from the end of his needle onto a stack of papers parked by the phone.

"There's your synthetic fiber," he said, turned on his heel, and walked out.

Within months of making this momentous discovery, Carothers sank into the worst bout of depression he had suffered since arriving at Du Pont six years before. As *Fortune* speculated:

> This sudden and profuse new world lost its interest for Dr. Carothers almost as soon as he had mapped it. He couldn't seem to interest himself in any new researches. He was tired and ailing. In nine short years he had envisaged hundreds of new compounds and tossed off no less than fifty patents. ... He had organized the basic field of polymerization out of chaos, supplied it with new techniques and a vocabulary, laid the groundwork for a revolutionary chemical manufacturing process. But after the fiber development, black depression set in.

At the height of his nervous collapse, he lost by his own account "all interest in chemistry." By February 1935, he had recovered sufficiently to supervise the creation of a superpolymer formed from an amine and an acid—hexamethylenediamine and adipic acid. The resulting polymer (polyhexamethylene adipamide) was formally denominated "Fiber 66" because both the amine and the acid contained six carbon chains.

In a final act of corporate humiliation, when Carothers humbly proposed that a different polyamide denominated Fiber 5–10 (made from pentamethylene diamine and sebacic acid) might be a superior candidate for development from a strictly scientific point of view, the ever practical Bolton arrogantly overruled him. The logic was inescapable: Benzene, a key ingredient in Fiber 66, was commercially available at a reasonable cost. Route 66 was cheaper by the dozen.

In early 1935, the development of synthetic silk was summarily wrested away from its hypersensitive inventor (by then saddled with a well-deserved reputation for being moody and mercurial) and turned over to the nuts-and-bolts boys. Du Pont president W. S. Carpenter Jr. issued strict instructions to the Experimental Station that "the time between the test tube and the counter" be reduced to an absolute minimum.

To that end, a vast force of chemical engineers was hastily transferred from the engineering and rayon departments, to start tackling the practical issues related to solving problems posed by mass production. Carothers and his staff at Purity Hall, officially hailed as conquering heroes, were left out of the loop.

Over the next two years, as Carothers's purpose at Du Pont dwindled to the point of insignificance, his bouts of depression became more frequent, more prolonged, more severe, and more intractable. Though not yet forty, he felt—and was made to feel—over the hill. According to his Philadelphia psychiatrist, he became irrationally obsessed at this point in his life with his abject failure as a scientist. Even his 1936 election to the National Academy of Science struck him as little more than a cruel mockery. After bluntly rejecting an offer to become chairman of the chemistry department at the University of Chicago on the preposterous grounds that his "nylon research had reached an exciting stage," in June 1936 he suffered a nervous breakdown from which he never fully recovered.

Involuntarily confined to the Philadelphia Institute, a private psychiatric facility, for five weeks, he wrote a melancholy missive to

Jack Johnson, his old friend, to whom he had once exultantly exclaimed, "Harvard was never like this!"

The above address is that of an especially elegant, large, and elaborate semi-beige house. It is so large that one only occasionally encounters a fellow patient. Most of them seem to be bright and cheery, very well-to-do kind of professional hypochondriacs. Meeting a doctor is a rare event. Treatment consists of conversation, rambling, inconsequential, pointless, and sometimes so repetitious and puerile as to be the source of laughter, amusement, or anger.

On the other hand, there are tennis courts, bowling, a pool of sorts, and a badminton court. Also hydrotherapy, which is a very elaborate and impressive method of taking baths. It must be terribly expensive, but they don't send bills until the patient's resistance has been fully built up (or broken down?). I am probably unconsciously (?) bankrupt now financially and certainly am so proving. Just too lazy to move.

This is an involuntary business so far as I am concerned. It was sprung on me suddenly about five weeks ago, and here for the most part I have been ever since. It's an all right place to loaf and since the diplomatic axe fell and I was transported up here I haven't been capable of anything but loafing anyway.

The future is pretty obscure. At the moment, the assignment seems permanent unless I simply walk out. They can't very well prohibit that here. So it is quite within the range of possibility that you may before long discover a disreputable caller at your front (or perhaps better back) door step. If so, don't shoot. It will be only me.

Just four months before, in a desperate bid to set himself back on the right track, on February 21, 1936, Carothers had married the lovely Helen Everett Sweetman of Wilmington, Delaware, a recent graduate of the University of Delaware. She and Carothers had met while she was working in the patent division of Du Pont's chemical department—during the time when he was working on his nylon patent. On January 8, 1937, a devastating blow struck too close to home: His beloved sister Isobel (who had become famous as "Lu" in a popular radio singing trio popularly known as "Clara, Lu, and Em") died suddenly after an illness of only three days. As his friend and mentor Roger Adams later observed, "Her death . . .

was a staggering loss to him. He was never able to reconcile himself completely to her loss."

In March 1937, Du Pont paid for a lavish recuperative trip to Europe, but the prescribed long vacation did not prove therapeutic, even if it was what the doctor had ordered. On April 9, 1937, Carothers filed for a patent application covering the basic chemistry of nylon. "Insofar as I am aware," he wrote, "the prior art on synthetic polyamide fibers, and on polyamides capable of being drawn into useful fibers, is nonexistent."

It should have been the culmination of a remarkable career. But it was not the career he had mapped out for himself when he joined Du Pont at Charles Stine's high-minded urging. He may have been naive in accepting Stine's repeated assurances that "pure research" would be all he would be asked to accomplish. He had advanced polymer chemistry beyond the theoretical scope of Staudinger's abstract speculations, into the realm of the real. But the product of his labors, under the pressure exerted by Benger and Bolton, had been perhaps *too* real: A fiber whose ultimate purpose would be to replace silk in stockings.

Nineteen days after filing his patent application, on April 28, 1937 (two days after his forty-first birthday), a distracted Carothers left his Wilmington home heading for Philadelphia, telling his wife that he was going away on "a short trip." At five the next morning, he checked into a hotel room in Philadelphia, from which, a few hours later, guests in the adjacent room reported hearing loud moans. When, just at the crack of dawn, the door was forced open by house detectives, Wallace Carothers was found dead on the floor. He had swallowed the contents of a vial of cyanide, into which he had squeezed the juice of a lemon, hoping to disguise its bitter taste.

7

The Vinyl Solution

Courtesy of The Henry Ford Museum & Greenfield Village

By his seventies in the 1930s, Henry Ford had become as obsolete as one of his early-model Tin Lizzies. Publication of his paranoid theories of a worldwide Jewish conspiracy in the *Dearborn Reporter* had led to a total Jewish boycott of Ford products. Pitched battles with organized labor and virulent opposition to the New Deal had saddled him with a fearsome reputation as a feudal despot. His staunch opposition to American involvement in the war against Hitler, and a dislike of Roosevelt "bordering on the psychopathic" (according to one biographer) all contributed to a widespread belief that Henry Ford was in steep decline—and taking the Ford Motor Company into the abyss with him.

The merest mention of American participation in World War II reduced him practically to incoherence. Yet as the world edged closer to a second global conflict, General Motors and Chrysler surged ahead of Ford in the marketplace. On his last legs, the grand old man of Detroit secretly prepared to pull one last ace from up his sleeve. In November 1940, the seventy-seven-year-old auto baron invited the press to Greenfield Village, his privately financed museum of Americana, to witness a demonstration of a radical new Ford.

As the photographers' bulbs flashed, the lanky old throwback swung a woodsman's ax with all his might into the trunk of a shiny new black sedan. Instead of denting, the glossy panel flexed dramatically, sending the heavy ax blade rebounding high into the air. The crowd of reporters and photographers standing in attendance jumped back a few paces, for fear that the old man might finally be losing his grip. But Ford was firmly in control of his faculties. Clutching his ax handle and beaming boyishly for the cameras, he slammed his ax into the trunk one last time for emphasis before brashly challenging everyone present to do the same to *their* cars. Pointing gleefully to the car's unblemished rear end, Ford formally announced: "We will make the entire superstructure of an automobile body, except the tubular welded steel frame, out of fiber plastic. The first experimental model will be finished this winter."

It was official: The disclosure that a plastic Ford was in the works sent shock waves of anticipation through a nation already threatened with acute metal shortages. On the brink of a second world war, it looked as if Henry Ford, in his dotage, had once again come up with the right product at the right time. The *New York Times* pronounced the plastic Ford "a great influence on the automobile industry." The Decatur *Herald Review* felt free to go even further overboard: "Here is something an America on wheels has been waiting for. Please hurry it, Mr. Ford! Hurry, hurry!"

The remarkable thing about the prototype plastic Ford was that it was made of a unique soybean-based resin that Ford and a young in-house plastics protégé had been secretly working on for twelve years. To Ford, a decades-old dream of "growing a car like a crop" had evolved into a full-scale campaign to reconcile two warring sides of his personality: the mechanical genius and the rural conservationist.

Born and bred on a farm at a time when farmers were routinely deprived of the benefits of modern technology, Ford had developed the first tractor—he called his Fordson an "automotive plow"—in an innovative bid to mechanize American agriculture. He had invented the Model T so that nongentleman farmers could negotiate the notoriously bad American country roads. And he had devised the assembly line to keep production costs low enough so that every farmer could afford to buy a basic black Ford to haul himself and his goods to and from town.

But in his mature years, Ford was appalled by the bleak, scarred industrial landscape the automobile had left everywhere in its triumphant wake. Longing to recapture and restore the bucolic small-town America that cars had done everything in their piston-power to destroy, he founded Greenfield Village, a tidy little theme park of preindustrial America, cobbled together from historic buildings plopped down in a Michigan cornfield. In its grave, naive simplicity, Greenfield Village reflected Ford's deep-seated desire to turn back the clock to the homespun America he had known as a boy on the farm.

During the twenties, and particularly during the depressed thirties, Ford had become a passionate proponent of "farm chemurgy," a theory developed by Dow Chemical chemist William J. Hale that held that the replacement of the horse by the internal combustion engine, which didn't eat cellulose and returned no useful fertilizer to the soil, had been an unmitigated disaster for agriculture. Fancying himself the farmer's friend, and feeling personally responsible for his plight, in 1929 Ford established a botanical laboratory at Greenfield Village to study plants with a view to unlocking their industrial potential. Hale had eloquently spoken of agriculture's sacred ability to create raw materials out of "the Three Musketeers of Nature": cellulose, vegetable oil, and alcohol. By 1931, Ford had isolated his own personal agro-industrial Holy Grail: the lowly soybean. As *Fortune* wryly commented, he was "as interested in the soybean as in the V-8."

The following year, Ford planted three hundred varieties of soybeans on eight thousand acres of his own prime Michigan farmland, and urging his fellow farmers to follow suit, personally guaranteed that he would find a market for them. By 1935, Ford made good on that promise: Inside every Ford car rolling off the line at River Rouge, a bushel of soybeans had been churned into oil to make enamel paint. Or ground into meal and molded into timing gears, horn buttons, gearshift knobs, door handles, accelerator pedals.

By his side every step of the maverick way was his handpicked counterpart to Wallace Hume Carothers—his homegrown chemical boy wonder, Robert Boyer. Unlike Carothers, who would go down in history as the preeminent scientist of the synthetic age, young Boyer's life's work was devoted to fulfilling his master's dream of forging a union of nature and artifice by developing an agro-plastic based on beans grown in the rich loam of rural Michigan.

Ford had discovered the fifteen-year-old Boyer in 1925, at Ford's Wayside Inn in Framingham, Massachusetts, another nostalgia-driven "historic" Ford restoration managed by Boyer's father. Entranced by the precocious teen's "active interest in what made the world go around," Ford plucked Boyer out of Framingham High School and enrolled him in the Ford Trade School, where hundreds of boys in whom the master ardently believed were trained to become adepts in the industrial arts.

At FTS, Boyer blossomed into a passionate devotee of the soybean cult. In 1930, as a graduation present for his private boy genius, Ford built for the twenty-year-old Boyer an old-fashioned three-story frame building located directly behind the museum in Greenfield Village. It looked like an old farmhouse. But behind its austere nineteenth-century facade lay a modern laboratory, lavishly outfitted with the very latest in polymer-processing equipment.

Comfortably ensconced in Ford's eccentric version of Du Pont's Purity Hall, Boyer (who lacked a formal degree in chemistry) was assisted by a team of twenty-eight scientists—twenty more than supported Carothers at Du Pont. Degree or no degree, Boyer devised innovative ways of extracting from soybeans an enamel for paint and a lubricant for casting molds. He developed a soybean-based "wool," which with its natural crimp and fine hand, struck the legume-besotted Ford as ideal for car upholstery.

Every morning before his daily round of the Rouge, Ford would drop in on Boyer in his plain, white-walled lab to observe the progress of chemurgy at work. "When his tan lean face bends next to Boyer's ruddy, cherubic face over the desk," *Time* keenly observed, "they look like two schoolboys plotting a prank." It was common knowledge around Ford that anything Boyer asked for, Boyer got—including a thousand-ton steel press the size of a ten-story building. Destined for delivery at the production department at River Rouge, Ford conspired with Boyer to divert it to Greenfield Village, where it was put to work heat-molding plastic panels for the secret soybean Ford.

At thirty, the stout, balding Boyer had acquired quite a few of his master's idiosyncrasies, from total abstinence from tobacco and alcohol to a professed fondness for the ballroom dances his mentor shamelessly adored. Aping Ford, Boyer ate soybean soup and soybean bread with conspicuous relish at every conceivable photo opportunity. The master, determined to carry his soybean fetish to ever greater extremes, rarely set foot in public without sporting a tie woven from Boyer's soybean "wool" or rigged out in a natty soybean suit, in which he appeared "as delighted as a little boy in his first pair of long pants," according to the *Detroit Times*.

On a warm Wednesday in August 1941, on the gala occasion of his annual Greenfield Day celebration, Ford proudly unveiled before a captive audience of ten thousand guests the fruit of more than a decade of covert labor: a cream-colored, blunt-nosed, soybean-plastic Ford sedan. With its standard chassis, engine, and wheels, it didn't depart visually from a conventional 1942 Ford. The difference lay hidden beneath its finely finished soybean skin, where an innovative tubular steel frame invisibly compensated for the plastic body's relative lack of structural strength. But the great difference between a plastic and a steel car was weight: A plastic car weighed in at two thousand pounds as opposed to steel's three thousand.

The public was intrigued and amused by the notion of a vegetable car. The *Cleveland Press* suggested that Ford consider enhancing its ingredients by adding a dollop of spinach. The *Cedar Rapids Gazette* coined a new Ford slogan: "Ask the man who *grows* one." Editorialists quipped that the new Ford didn't need gas—just sprinkle a little salt, pepper, and vinegar on it, and it would run like a top. Instead of buying a new vegetable car every year, an owner could just have his old one warmed over. And the lucky man who owned one of the new vegetable Fords could always "have his car and eat it too."

Yet even with the best press in the world, Ford's timing could not have been worse. Though William S. Knudsen, chairman of the National Defense Advisory Committee, expressed interest in any car that promised to free up tons of scarce steel during the ongoing "national emergency," any remaining chances of the plastic Ford hitting the road soon were decisively dashed at dawn on Sunday morning, December 7, 1941.

Pearl Harbor

On that sleepy Sunday, President Roosevelt sent word to his wife, Eleanor, that he would not be joining her and a few guests at a formal lunch but would instead dine casually with his friend and adviser Harry Hopkins in his office on the second floor of the White House.

At 1:40 P.M., just as the steward was discreetly clearing the plates, Navy Secretary Frank Knox called on the president's private line, eager to relay an urgent message just in from CINCPAC, Naval HQ, Honolulu:

"Air Raid Pearl Harbor. This is no drill."

Later that day, Roosevelt's son Elliot, then serving as an army air force officer at Kelly Field, Texas, rang up to ask the probing question: "What's the dope, Pop?" The dope was that the Japanese had launched a devastating sneak attack on Pearl Harbor, rapidly reducing the once mighty U.S. Pacific Fleet to fewer than two dozen hulks of smoldering rubble in under two hours.

Preparing for the emergency joint session of Congress at which he planned to ask Congress to formally declare war on Japan, FDR reviewed the first draft of his proposed speech. The phrase "a day that will live in history" caught his eye. He drew his pencil point through "history" and neatly wrote in "infamy."

Rubber Realities

On that otherwise sleepy Sunday afternoon, Sunday drivers across the country, by force of friction alone, left over five thousand tons of rubber on the surface of the nation's highways. By 1941, the United States had become the largest consumer of crude rubber in the world, annually devouring two-thirds of the annual global output—over half a million tons per year. With 80 percent of the cars in the world spinning their wheels on American roads, it was hardly surprising that U.S. rubber consumption ran six times that of Great Britain, ten times that of Japan, and twenty times that of the Soviet Union.

So long as the "impregnable" British fortress island of Singapore remained in friendly hands, that the nation's rubber supply line was ten thousand miles long was of no great concern to American consumers or military strategists. What did it matter that over 90 percent of American crude rubber came from two East Asian sources—Malaysia and the Dutch East Indies? But despite repeated assurances on the part of British and U.S. military leaders that Singapore would never fall to enemy attack, the day after Pearl Harbor the Japanese landed in Malaysia and advanced steadily toward Singapore, guardian of Java and Sumatra—heart of the world's rubber belt.

It took the Japanese another two months to take Singapore, but as the Axis grip tightened on America's only reliable source of raw rubber supplies, the country was plunged into the only raw materials crisis it had ever known. America was an astoundingly resource-rich country, except for one strategically critical commodity: rubber. As modern war increasingly relied on rubber tires to move every tank, plane, truck, and tractor-trailer in its strategic arsenal, a lack of homegrown rubber became America's looming Achilles' heel.

Japanese domination of America's East Asian rubber supply line turned out to be of far greater consequence to the long-term military security of the United States than the devastating loss of the Pacific Fleet at Pearl Harbor. While Fortress America could build more ships and planes, no plane could land, no truck or jeep move down a road, without rubber tires to cushion its path. In the dark days following Pearl Harbor, for all the tough talk about forging an "arsenal of democracy" for America's allies, the little acknowledged (for fear that it might sow panic in an already tense population) weak link in the armor of Fortress America was rubber. Or rather, the lack of it.

As Donald Nelson, chairman of the War Production Board, later divulged:

> Rubber caused more serious headaches in the defense and war-production agencies than any other single material. . . . We knew we would have to fight a highly motorized and mechanized war or a losing war. We knew we could not mount a highly motorized, mechanized war [without] rubber.

Two days after Pearl Harbor, the Office of Production Management (OPM) banned sales of rubber tires and tubes to civilians, with the exception of doctors and operators of essential commercial vehicles. The OPM also launched an intensive public-information campaign to urge all Americans to turn in their discarded tires and worn rubber bath mats in the most strategically critical "scrap drive" of the war. But over the next three years, despite vast expenditures of money and effort on the part of the Rubber Reserve Company (the quasi-government agency set up to safeguard the nation's dwindling rubber supplies) just a hundred thousand tons of raw rubber were imported from two locations: Liberia and Brazil. In terms of equipping the largest expeditionary force the world had ever known, "reclaimed rubber" provided hardly even a drop in the bucket.

As the war escalated, the conflict between the Axis and Allied powers became a race against time for control over a wide range of strategically critical materials. Wherever nature was deficient, chemists were called upon to make up the difference. The earliest and most critical contest between the Axis and Allied powers took place in the realm of rubber, both natural and synthetic.

The roots of the current crisis dated back to 1934, when the British and Dutch rubber planters in the Far East formed the International Rubber Regulation Committee, which openly dedicated itself "to adjusting in an orderly manner supply to demand." In less euphemistic terms, the goal of this cartel was to squeeze the world's supply of rubber to force up the price. While the depression slashed raw rubber prices down to an all-time low of four cents a pound, once the Rubber Regulation Committee got into gear, rubber prices returned to a more respectable level—twenty or so cents a pound.

Even at those prices, synthetic rubber, only just hitting the market, could never have competed with the real thing. While a complacent U.S. military establishment paid little heed to developing a domestic synthetic rubber capacity, the Germans, apt students of the lessons their materials-strapped nation had been forced to learn during the last war, adopted more strenuous measures.

Under the Nazis, Hermann Göring's Ministry of Economics declared the manufacture of "ersatz caoutchouc" the number one research priority of the *Reichsforschungsrat*—the government-coordinated research effort to put science at the service of the state.

As Reichsminister Bernhard Rust, chief of Nazi learning and training, chillingly proclaimed in a prewar address to a national scientific assembly:

> The Nazi state calls upon German science to cooperate in the Four Year Plan. . . . It is the task of science to give us those materials Nature has cruelly denied us. . . . Science is only free if she can sovereignly master those problems posed to her by life!

Under Hitler's tightly controlled Four Year Plan, German tire makers were required to use 10 percent Buna synthetic rubber for all tires manufactured in the first year (1935), with the quota rising to 25 percent in the second year, and so on up until 1939, when Nazi tanks and trucks were expected to conquer vast new territories on a Buna-cushioned *Blitzkrieg*.

As Hitler exultantly decreed to a roaring crowd at a Nuremberg rally on the eve of the invasion of Poland, "We have definitely solved the rubber problem!" This was in part Hitler bluster and part objective truth: Germany had gone farther down the synthetic road to off-setting its natural deficiencies than any of its future adversaries.

Gumming Up the Works

That crude rubber is a derivative of the chemical compound isoprene was discovered by the British chemist Greville Williams in 1860. A quarter-century later, Sir William Tilden, another British chemist, prepared isoprene from turpentine and left it standing in a clear glass bottle on a shelf in his laboratory. After absorbing strong sunlight pouring in from a nearby window for a few days, no one could have been more surprised than Tilden when, after opening the stock bottle, he found that in the heat the liquid had decomposed into a viscous, sticky, transparent gum. At first Tilden assumed that he had stumbled upon synthetic rubber. But he had merely advanced the process a crucial first step: turning isoprene gum into a polymer.

It took the crack team of chemists assembled by Germany's I.G. Farben to discover a copolymer (a polymer made of two compounds polymerized together) of butadiene and styrene that yielded a viable synthetic rubber. They called their new elastomer (elastic polymer) Buna—a name that fused the *Bu* in butadiene with Na, the chemical symbol for sodium, and the catalyst for the polymerization reaction.

While Germany, lacking petroleum supplies, had relied on coal-tar derivatives to produce its synthetic rubber, in April 1931, Wallace Carothers and his team at Du Pont discovered Neoprene, a polymer of chloroprene—a petroleum derivative. Two years before, while testing a compound of ethylene chloride and sodium polysulfate for its suitability as an antifreeze additive for gasoline, two independent researchers, J. C. Patrick and Nathan Mnookin of Kansas City, had succeeded in fabricating America's second answer to German Buna. Their intractable gum gave off a vile rotten-egg sulfuric smell and displayed a remarkable capacity to clog a drain in a lab sink. Still, the stuff looked and acted remarkably like rubber, with a few notable exceptions: No known solvents dissolved it. Plus—possibly a very big plus—it was oil and gas resistant.

Convinced that the new rubber would make an unusually durable hose for oil and gas pumps, Bevis Longstreth, owner of the Morton Salt Company, put up enough money to construct a small

plant in Kansas City to produce "Thiokol"—a name coined by
Mnookin and Patrick by combining the Greek roots for "rubber"
and "sulfur." But with Thiokol selling—or not selling—at sixty
cents a pound, and with Du Pont's Neoprene still pegged at a buck
a pound, domestic rubber producers had little incentive to invest in
synthetics.

Not, that is, until Pearl Harbor. Following Singapore's shocking
collapse, with strategic concerns suddenly paramount, price became
no object. Unfortunately, politics did. With congressional interest in
synthetic rubber fueled by the prospect of juicy government contracts, various regional factions and interest groups began to con-
tend for government backing.

Butyl, a petroleum-based rubber developed by Standard Oil,
was the clear favorite of petroleum administrator Harold Ickes, and
of the politically powerful petroleum industry. But butyl's bid was
blocked in Congress by the powerful Sen. Guy Gillette of Iowa,
who preferred his synthetic rubber made from alcohol distilled from
wheat and corn. The California congressional delegation, meanwhile, pulled every string in the book on behalf of an experimental
rubber made from the ubiquitous rabbit bush, after Dr. T. Harper
Goodspeed of the University of California testified before Congress
that the rabbit bush could readily be harvested on the 3.7 million
acres of government-owned scrubland located in—where else?—the
great state of California. Rep. August Andresen of Minnesota, a
fierce partisan of a rubber fabricated from the Russian dandelion,
urgently requested Secretary of State Cordell Hull to ask the Soviets
to send ten tons of dandelion seeds to start up a dandelion-rubber
plant to be situated in, of all places, northern Minnesota.

With politicians and lobbyists all squabbling and scrambling for
leverage, no progress could be made on advancing the nation's synthetic rubber capacity. While ordinary citizens were being called
upon to fill in the gaps by rummaging though backyards and
garages for old rubber tires, even the most loyal Americans could
hardly be expected to part with old tires without any hope of getting new ones.

Suggestions for curing America's rubber crisis occasionally bor-
dered on the inane, if not insane. A Los Angeles dentist gained fleeting fame by calling on highway authorities—in a syndicated radio
broadcast—to scrape the tons of waste rubber off the nation's roads

and highways, "particularly at hard stops and sharp curves." At a meeting of government officials seeking to tackle the crisis, Leon Henderson, an official with the Office of Price Administration, suggested that the average length of latex condoms be reduced by half. It took a moment of silence for the assembled participants to grasp that he was kidding.

By mid-1942, with the strategic rubber preserve hitting all-time lows, a furious President Roosevelt appointed the legendary financier Bernard Baruch—who had served as civilian coordinator of defense production during World War I—to join Harvard president James B. Conant and MIT president Karl T. Compton on a fact-finding panel charged with getting to the bottom of the rubber crisis. Their scathing report, issued in August 1942, concluded on a note of distress:

> Of all the greatest critical and strategic materials, rubber presents the greatest threat to the safety of our nation and the success of the Allied cause. Production of steel, copper, aluminum, alloys, or even aviation gasoline may be inadequate to prosecute the war as rapidly and effectively as we would wish. At the worst, we are still assured of sufficient supplies of these materials to operate our armed forces on a very powerful scale. But if we fail to secure quickly a large new rubber supply, our war effort and our domestic economy both will collapse.
>
> To dissipate our remaining stocks of rubber is to destroy one of our chief weapons of war. . . . The country is dependent upon the production of synthetic rubber, which it is hoped will reach full swing in 1944. Why not earlier? Why so late? These errors grew out of procrastination, clashes of personalities, lack of understanding, delays and the early non-use of known alcohol processes. . . .

And, in the shocking case of one influential U.S. company, fraternizing with the enemy verging on treason.

I. G. Moloch

During the Nazis' rise to power in Germany, the staunchly "internationalist" character of the chemical colossus I.G. Farben—combined with a penchant for appointing Jewish scientists to senior positions—made it the frequent target of crude lampoons in Nazi Party publications, replete with snide references to "Isidor G. Farber"—accompanied by a grotesque caricature of Shylock—and "I. G. Moloch," a mocking reference to the Canaanite God to whom children were ritually sacrificed.

But by 1932, a once notoriously progressive and tolerant company had embarked down the road to transforming itself into a vehicle of Nazi aggression. On the night of February 20, immediately following Hindenburg's reluctant elevation of Hitler to the chancellorship of Germany, Farben director Baron Georg von Schnitzler attended a secret meeting of top industrialists and financiers held at the home of Nazi Hermann Göring. Rising to make a dramatic pledge on behalf of the Farben companies, Schnitzler promised to raise four hundred thousand of the estimated three million marks the Nazis would need to gain decisive control of the Reichstag in the upcoming elections.

After Hitler succeeded in maneuvering himself into a position of absolute power in Germany, Farben's new chairman Carl Bosch—Nobel Prize–winning chemist Fritz Haber's partner in the synthesis of ammonia—respectfully requested an audience with him to seek support for his company's synthetic oil project. In incalculable debt to his number one corporate sponsor, Hitler pledged the new government's unwavering assistance. This came as no great surprise to Bosch, since Hitler was known to be as fanatic a proponent of German self-sufficiency as he was an avowed anti-Semite.

But when Bosch, encouraged by Hitler's conciliatory tone, seized this prime opportunity to warn the Führer that if Jewish chemists and physicists were deported from Germany, such a step would set German physics and chemistry back a hundred years, Hitler stood and roared so the chandeliers shook:

"Then we will work a hundred years *without* physics and chemistry!" Ringing a little silver bell on his desk to signal that the meeting was over, Hitler stormed out of the room in disgust, never to been seen in the same room with Bosch again.

Two years later, Fritz Haber—a staunch German patriot much maligned for his role as head of Germany's brutal poison gas effort during the last war—was forced to resign his professorship at Berlin University and face summary deportation to Switzerland because, though a convert to Christianity, he was Jewish born. Bosch and other I.G. Farben officials tried to intercede on his behalf, but Haber was duly exiled to Basel, where he died six months later, a broken and bitter man. When the Nazis prohibited German citizens from mourning his passing, Bosch defied the ban by organizing a memorial service in Haber's honor. Clearly, under the new dispensation, the days of the Jew-lover Bosch at the head of Germany's largest chemical cartel were numbered.

By 1937, the second year of Hitler's vaunted Four Year Plan, I.G. Farben had transformed itself into the model of the modern Fascist conglomerate—Aryanized, Nazified, eager and willing to serve the Third Reich. Board members who had managed to evade joining the party until then either signed up or resigned in disgrace. Jewish officials and employees, including a third of the advisory board, were deprived of titles, offices, and pensions. After Hitler's secret attack on Czechoslovakia on October 1, 1938, Bosch was quietly removed from his post. With the rousing slogan "Guns before butter!" the freshly appointed economics plenipotentiary Hermann Göring grimly ordered the German people to prepare for total war.

By late 1938, armaments minister Albert Speer was sardonically referring to the Four Year Plan as "The I.G. Plan." This was hardly an exaggeration: The combined Farben companies were by then manufacturing 85 percent of Germany's strategic materials under the supervision of Göring's all-embracing Economics Ministry.

As the possibility of war in Europe threatened to bring the United States and Germany into conflict, the Nazis and I.G. Farben skillfully connived to thwart the development of an independent synthetic rubber capacity in the United States. The Nazis were particularly concerned that Standard Oil of New Jersey, which had enjoyed a close commercial relationship with I.G. Farben for ten years, might gain access through that relationship to Buna rubber— the most versatile of all the synthetics. Nothing about the tightening alliance between the Nazis and Farben disturbed Standard Oil's friendship with Farben. It was a relationship dating back to 1930, when Standard Oil chairman Walter C. Teagle had moved aggressively to lock up U.S. rights to Farben's top-secret coal-to-gas process for fear that one day it might undermine the entire American petroleum industry. Later that year, Farben and Standard had cofounded an American joint venture, Jasco, to serve as a corporate repository for any patents Farben and Standard might wish to swap in future deals.

For the remainder of the decade, if Farben chemists cooked up a new compound that might be of interest to Standard, Standard would have its chemists and marketing people check it out for possible purchase of the U.S. rights. The same went for Standard discoveries, like the petroleum-derived butyl rubber. Just as Hitler and Stalin would one day slice up Poland, the Standard and Farben cartels neatly divided the world between two noncompeting "spheres of influence."

At a March 1938 meeting between Dr. Fritz ter Meer, chief of Farben's Technical Committee, and Standard Oil vice president Frank Howard—held just days before Hitler's annexation of Austria—ter Meer persuaded Howard to agree to hand over Standard Oil's patents and technical information on butyl rubber (Standard Oil's petroleum-based synthetic rubber) in exchange for a *promise* on ter Meer's part that Farben would try to secure permission from the Nazis to hand over equivalent information on Buna—Farben's styrene-based synthetic rubber.

After shaking hands on this spectacularly uneven swap—a patent for a promise—Howard sent a memo to his superiors urging them to take Dr. ter Meer's word as his bond.

> Considering the very genuine spirit of cooperation which Dr. ter Meer displayed, I am not only convinced that it is the right thing to do but the best thing from every standpoint to pass on to them full information on [butyl] at this time. . . . I do not believe we have anything to lose by this.

What Howard did not know at the time would in due course destroy him: Ter Meer had reassured Brigadier General Loeb of the Four Year Plan that he had no intention of supplying Standard, or any other American company, with a license to make Buna—certainly not without Loeb's permission. The true object of his meeting with Howard, he fervently insisted, had been "to halt the development in the U.S.A. of a synthetic rubber capacity."

In support of this devious plan to sidetrack the United States' synthetic rubber problem before it began, Dr. ter Meer scheduled a series of sham meetings with Dow Chemical, Goodyear, and

Goodrich representatives (all champing at the bit to buy Buna) with, he told Loeb, "the sole object of easing the minds of American interested parties, and *to prevent any independent initiative on their own part.*"

Noting that Goodyear had "made some progress in copying our Buna," ter Meer advised Loeb that "because tendencies for restoring military power are very strong [in America] too ... we shall have to treat license requests by American firms in a dilatory fashion, *so as not to push them into taking unpleasant measures.*"

By "unpleasant measures," ter Meer was referring to American attempts to develop a synthetic rubber program outside the boundaries of the I.G.-Standard accord. And by "dilatory fashion," he was plainly referring to the skilled stalling by which he was forcing Howard to bide his time waiting for the green light from Berlin.

In the meantime, Howard was holding his own series of bogus conferences with Goodrich, Goodyear, and Dow, at which he intentionally misled them into believing that permission to sublicense Buna was on the verge of being granted. To Chairman Walter Teagle and President William Farish, Howard was a little more candid: "My object was to convince them of our good faith and our willingness to cooperate with them *in order to avoid having them proceed prematurely with an independent development.*" Howard and ter Meer were not just employing the same tactics, they were using the same phrases.

September 1, 1939: As scores of Nazi panzer divisions rumbled into Poland on Buna-tread tires, Frank Howard cabled ter Meer from France, urgently requesting a meeting to be held at The Hague in neutral Holland. Before proceeding, Standard checked with its contact in London, American ambassador Joseph P. Kennedy (a notorious appeaser of German interests), who evidently regarded any sign of German-American cooperation at this late date as a positive development.

At The Hague, with the blessing of Ambassador Kennedy, Howard and ter Meer hastily hashed out a strategy for dealing with the worst-case scenario: that America and Germany would shortly be enemies, and that in the event of open hostilities between the two countries, Standard's Buna patents would be subject to forfeiture as a transfer of "enemy property" by an alien property custodian.

As a means of keeping valuable trade secrets out of the clutches of "unfriendly interests," ter Meer proposed turning the patents over to Jasco, the U.S. repository for pooled Standard-Farben patents. Back in Berlin, after being repeatedly warned by ter Meer that if he didn't let Howard have his Buna patents, Standard and the other U.S. companies would be forced to "go it alone," General Loeb grudgingly granted Farben permission to release the patents to Howard. But on one condition: that there be no transfer of technical knowledge, and that the patents, and the patents alone, be all that slipped out of their grasp. Without the underlying technical material, the patents would not provide, in the opinion of Loeb and ter Meer, a detailed enough blueprint for Standard to commence Buna production.

In a cable to Howard dated October 12, 1939, ter Meer formally notified Standard of Farben's willingness to assign Farben's Buna patents to Jasco. But—and this was a *big* but—ter Meer signed off: "In reference to your request for technical information about Buna, we regret to inform you that under present conditions we will not be able to give such information." Only then did Howard realize that he had been the victim of a classic Nazi deception.

In the wake of Pearl Harbor, and with awkward questions being asked about why the U.S. chemical industry had so badly botched the rubber question, the Antitrust Division of the Justice Department belatedly swung into action. It took the white hats and gumshoes at Justice two months of snooping around before they felt confident enough to slap Howard, Teagle, and Farish with criminal indictments charging them with restraint of trade in two essential strategic commodities: synthetic fuel and synthetic rubber.

Standard Oil's attorneys were shockingly successful in mounting a vigorous and effective pretrial defense on behalf of their beleaguered client. Regardless of the merit of the charges, the lawyers argued that for Standard Oil to be forced to defend itself from an antitrust suit at such a critical juncture in the nation's history would severely hamper its ability to engage in the war effort.

This was out-and-out blackmail. But it worked like a charm. The War Department, more eager to facilitate defense production than belatedly enforce moral judgments, took Standard's side in the ensuing power struggle. In a decision worthy of Solomon, President

Roosevelt, while agreeing in principle to call off the dogs on antitrust suits that threatened to disrupt the war effort, insisted that Standard's unholy alliance with Farben was too egregious an assault on the American free market—and on American strategic interests—to be simply ignored.

In a halfhearted compromise, Standard was permitted to plead *nolo contendere* (no contest) to all government charges while deftly hoisting itself off its own petard by signing a consent decree pledging to abandon all contacts and practices with the enemy concern—though only for the duration, of course. Not that severing ties to a notorious Nazi conglomerate constituted any great sacrifice at that point. While an irate attorney general Thurman Arnold vowed to stick Standard with a settlement tab in the staggering range of $1.5 million, Standard attorney John W. Davis objected on the spurious grounds that such a settlement would "call Standard Oil's patriotism into question." Standard ended up being let off with barely a slap on the wrist for selling out America—a $50,000 fine for each of the five indicted Standard companies, for a grand total of $250,000.

In April 1942, Missouri senator Harry S. Truman, in his capacity as chairman of the Senate's War Investigatory Committee, conducted full-scale congressional hearings into the rubber scandal. The dramatic high point was Truman's chilling production of a "smoking gun memo" from Frank Howard to his superiors dated October 12, 1939—the day of his meeting with ter Meer at The Hague. In that memo, Howard calmly advised Teagle and Farish that "we [ter Meer and Howard] did our best to work out complete plans for a *modus vivendi* whether or not the U.S. comes into the war."

The meaning of this last sentence could not have been clearer: In the tragic event that such old friends as America and Germany went to war, the sacred, secret bond with Farben would come first. On March 25, 1942, the newly appointed U.S. alien property custodian, just as Howard and ter Meer had feared, seized the interests of I.G. Farben, "an enemy corporation"—including Jasco's prized Buna patents.

In the progressive paper *PM* (published by eccentric supermarket heir Huntington Hartford) liberal columnist I. F. Stone called upon John D. Rockefeller to "personally remove Walter C. Teagle as

Chairman of the Board and William S. Farish as President and Frank A. Howard as Vice President of the Standard Oil Company, and to radically change the policies which put them in the position of acting as international economic collaborators of the Third Reich."

But Mr. Rockefeller was prepared to do no such thing, certainly not at the behest of a radical like Stone. Howard remained at Standard for two more years, though in a state of disgrace, discreetly deprived of his former influence. After a November 1942 Roper poll found that "most Americans believed that Standard had let Germany best it in business," Teagle resigned from Standard's board. Less than a week later, Farish died of a heart attack.

Fast Track

Once the contending parties were prodded into action by the threat of scandal and the prospect of imminent disgrace—if not outright prosecution—the American synthetic rubber crash program, following an excruciatingly slow start, took off like a rocket. An unparalleled degree of coordination and cooperation was ultimately achieved between academic and corporate labs, university professors and industrial chemists, and chemical companies that up until then had been fierce competitors.

The construction of fifty plants at breakneck speed at sites scattered from Connecticut to Louisiana to Texas was a monumental undertaking in itself. It required the allocation of over a million tons of scarce structural steel before a $1 billion industry (launched by the government from scratch) could be operated by a vast network of managers, supervisors, contractors, and subcontractors, all laboring under pressure to meet a strict congressional mandate that synthetic rubber production reach a level of eight hundred thousand tons by 1944. Midway through the target year, what had once seemed an unobtainable goal was not only reached but surpassed. By the war's end, total production of synthetic rubber had soared to a level far exceeding the highest annual output of *crude* rubber ever processed by the United States before the war.

As BFGoodrich president John Lyon Collier later proudly recalled: "When BFGoodrich was founded in 1870, United States rubber consumption was four thousand tons. In five years, from 1939 to 1944, the production of synthetic rubber rose from 1,750 to

800,000 tons. The synthetic rubber producers achieved in five years what the natural rubber industry took seventy years to accomplish."

Farben's Folly

Under unbearable pressure to ratchet up its own Buna production, I.G. Farben was forced to adopt more radical measures. Despite Hitler's hyperbolic insistence at Nuremberg that Germany had "definitely solved the rubber problem," the truth was more complex. If anything, the Führer had willfully compounded his country's natural material deficits by turning a deaf ear to the warnings of his general staff and industrial advisers that waging "total" war on two fronts would stretch German supply lines to the breaking point— particularly with regard to scarce oil and rubber resources.

Despite the unnerving fact that a month of intensive Luftwaffe bombardment had left the British Isles "an unsinkable aircraft carrier pointed directly at the heart of Germany," Hitler stubbornly refused to postpone his planned invasion of the Soviet Union by more than two months. Instead, as a means of offsetting the anticipated drain of a dwindling pool of raw materials, three months prior to Germany's scheduled invasion of the Soviet Union, Dr. Fritz ter Meer was summoned to a high-priority conference at the Ministry of Economics in Berlin to explore ways of increasing Buna production.

The solution to Germany's rubber shortage, ter Meer argued, was to construct a third Buna plant somewhere in the East to supplement the supplies being produced at Piesteritz and Huls. With his two plants inside Germany considered overly vulnerable to Allied attack, ter Meer proposed that a third plant be sited in one of the occupied territories—perhaps Poland or Czechoslovakia.

The prime moving factor behind Hitler's celebrated "Drive to the East" had been the ancient urge to conquer vast, as yet untapped reserves of raw materials, land, and labor for the benefit of the Aryan race. In strict accordance with the famed Nazi imperative of *Lebensraum* (making room), Dr. ter Meer dispatched one of his most talented synthetic rubber chemists, Dr. Otto Ambros (who eight years earlier had developed the theoretical basis for magnetic tape technology), to scour the recently occupied territories to the East in search of an ideal Buna site.

While on an inspection tour of Upper Silesia in southwestern Poland, on the border of Czechoslovakia, Ambros came upon "a chemist's dream." On the outskirts of Oswiecim, a town still too small to be marked on most Polish maps, the SS had constructed a concentration camp designed to house some three thousand inmates: mainly undesirable German Jews and political suspects, casually mixed in with a few hundred recalcitrant Poles and Gypsies. The good news was that the SS had announced plans to expand the camp (known in German as "Auschwitz") to accommodate thousands of Jews still imprisoned in the crowded ghettos of Warsaw and Lodz. Most were scheduled for imminent transport to the fabled "work camps"—not yet called "death camps"—in Poland.

"Nature has endowed this place!" Ambros exultantly wrote in a confidential memo to ter Meer, concluding with a glowing inventory of the region's material treasures: The tiny town straddled the southern border of the vast Silesian coalfields, a critical resource, given the official Farben estimate that the new Buna plant would in time consume in excess of a million tons of raw coal a year. Official estimates projected electrical consumption to ultimately exceed that of the city of Berlin. Much of this power, in a pinch, could be generated by damming one or more of the three free-flowing rivers—the Sola, the Przemsze, and the Little Vistula—that converged a short distance away.

Complementing nature's bounty was a highly favorable "human situation." Not only did thousands of farmers reside in the area who might be enticed into industrial work at decent wages, but "the availability of the inmates of the camp," Ambros advised, "could hardly be more advantageous."

Acting on the strong recommendation of Dr. Ambros, Farben's *Vorstand* (board) agreed to locate the new Buna plant on the outskirts of Oswiecim. With typical Teutonic efficiency, Farben set up an autonomous corporate division known as "I.G. Auschwitz" to handle construction and operations. The board became so completely convinced of the new division's long-term commercial prospects that it voted to underwrite the entire cost of construction—nine hundred million reichsmarks, worth $250 million—without resorting to a government subsidy.

In return for injecting this massive capital infusion into the subject territories, Farben obtained from the high command the highest possible construction priority. On February 18, 1941, Hermann Göring personally wrote to Hitler requesting that "the largest possible number of skilled and unskilled construction workers be made available from the adjoining concentration camp for the construction of the Buna plant." Since enslaved concentration camp inmates were estimated to be only 75 percent as effective as free German labor, the official allotment of slave labor battalions was naturally adjusted accordingly.

The greatest obstacle encountered by Farben in its fast-track construction program was not, as in the United States, the procurement of sufficient structural steel. Ter Meer and Ambros encountered their toughest times recruiting enough capos—inmate-foremen selected for their sadistic tendencies—to oversee the forced-labor details. On an inspection tour of I.G. Auschwitz in April 1941, SS Reichsminister Heinrich Himmler gave Otto Ambros his personal guarantee that the SS would supply at least ten thousand Auschwitz inmates to serve on construction detail. As Ambros excitedly cabled ter Meer, "Our new friendship with the SS is proving *exceedingly* profitable."

Unfortunately for the development of I.G. Auschwitz, the harsh methods the SS employed to compel starving slave labor to produce at maximum levels of efficiency were as often as not counterproductive. As a Farben on-site manager complained to his superiors in Berlin, the sight of "inmates being severely flogged at the construction site by Capos" seemed to have a demoralizing effect on the entire crew of free labor. In light of this negative influence, the Farben manager respectfully asked his superiors to ask the SS to order the capos to "refrain from conducting floggings at the construction site itself." Such floggings would be better off transferred, in his opinion, "to the camp itself," to minimize their demoralizing effect.

Floggings or no floggings, SS projections of the productivity of starving slave labor turned out to have been hopelessly optimistic. Even the deaths of some twenty-five thousand slave laborers during the construction of the two plants could not save Farben's one-billion-plus reichsmark investment in I.G. Auschwitz from coming perilously close to being written off. In July 1942, Farben's *Vorstand* voted to solve its labor problems at Auschwitz by establishing a private concentration camp, to be called Monowitz, at an adjacent site. When Camp Monowitz was completed in the summer of 1942—with the chilling motto *"Arbeit Macht Frei"* (Freedom Through Work) emblazoned over its arched wrought-iron entrance—the deadly imperatives of the Final Solution had long since superseded any remaining drive to solve Germany's raw-materials shortages. By the time Auschwitz was liberated in 1945, I.G. Auschwitz had contributed no more than a trickle of synthetic fuel to the German war machine, and not a single pound of Buna.

Radar War

Tuesday evening, August 23, 1940: as Winston Churchill pondered how Great Britain's tiny Royal Air Force could cope with the huge German air attack being launched from bases in France and Norway, some sixty miles south of 10 Downing Street RAF Flight Officer Ashfield of the new RAF Flight Interception Unit eased forward the throttle of his Bristol Blenheim, rising gracefully into a clear moonlit sky over Tangmere Aerodrome, near Chichester.

Banking slowly left over the blacked-out city, Ashfield nosed his craft eastward over the English Channel, spying a German Dornier 17Z in his sights. Firing his nose-mounted cannon at the target, he scored a direct hit. One Dornier down, 4,300 to go.

The top-secret weapon used by Officer Ashfield to intercept that Dornier had been secretly described by Winston Churchill as "the most elaborate instrument of war . . . the like of which existed nowhere else in the world." Though the technology itself remained shrouded in secrecy, British possession of a highly effective RDF (radar direction finding) capability was not in doubt. The jealous Germans had their highly accurate X-Gerät radio-beam navigational system, but it was not and never would be radar.

The ironic element in this saga of surveillance was that the British might never have developed radar if Adolf Hitler had not gloated in one of his radio addresses of the Nazis' possession of a "secret weapon" capable of shooting down enemy aircraft at great distances. With Buck Rogers–style death rays a staple of depression-era popular literature and comic books, it was impossible for the British Air Ministry to discount the possibility that German scientists had in fact come into possession of some form of death ray.

On a blustery morning in January 1935, Harry E. Wimperis, director of scientific research for the British Air Ministry, called Robert Wattson-Watt, head of the government's Radio Research

Laboratory at Slough, into his office for a strictly confidential man-to-man chat. Would it be possible, Wimperis wondered, to direct a beam of electronic energy at an airborne target, even at a great distance, powerful enough to "boil the pilot's blood"?

After mulling this macabre question over for a few minutes, Wattson-Watt, a canny Scotsman, candidly pointed out that even if such a death ray could be devised—which he doubted—unless the defending ground forces possessed the ability to track incoming aircraft with complete precision, even the most powerful death ray would be worthless as a means of defense.

Returning to his office at Slough, Dr. Wattson-Watt, who as a student in Brechin, Scotland, had developed an interest in radio telegraphy, spent the next couple of days pondering the possibility of directing enough electronic energy at a passing airplane to boil a pilot's blood. He soon realized that boiling blood would hardly be necessary, since elevating internal body temperature above 105 degrees Fahrenheit would cause the pilot to run such a high fever that he would quickly lose consciousness and perish in the ensuing crash.

Wattson-Watt felt confident advising Wimperis that Herr Hitler was, as usual, bluffing about the Nazi death ray. But he was sufficiently intrigued by the long-range implications of Wimperis's question that he shared his concerns with his assistant, Dr. Arnold "Skip" Wilkins. Wattson-Watt was struck by Wilkins's offhanded comment that a recent British Post Office report had noted disturbances to VHF reception when aircraft overflew powerful radio receivers. This curious phenomenon, Wilkins suggested, might prove useful in detecting enemy aircraft, since the radio signals interrupted were apparently reradiated by the passing aircraft groundward. Theoretically, it should be possible to detect aircraft with a series of sensitive ground-based receivers mounted on high towers. Wattson-Watt had in fact patented a concept he called *echolocation*—the science of locating objects by means of echoes. He had extensively studied the use of radio waves to track approaching thunderstorms, an essential task in aiding the safety of fragile aircraft.

The British Air Ministry was so impressed by Wattson-Watt's theory (and by the demonstration that he and Wilkins subsequently conducted for their benefit at an airfield just north of London) that they assigned him and a support staff of radio technicians to a top-secret RDF team unobtrusively headquartered at the Crown & Castle Inn in the scenic village of Orford, in Surrey.

The comprehensive air-defense scheme developed by Wattson-Watt required the construction of a chain of radar stations encircling the coast of Great Britain. But Wattson-Watt and his colleagues quickly concluded that even if the coastal chain succeeded in "deterring the day bomber . . . that a major consequence of fair-weather success . . . [would be for] the enemy to be driven to night attack, or to its near equivalent—attack through cloud."

The only reliable way of detecting attacking aircraft at night or through cloud that Wattson-Watt could conceive of would be to somehow "compress the radar station into an item of aircraft equipment." But with the only radar equipment then in existence (1936) "weighing many tons, requiring a power supply of public-utility magnitude, and aerials mounted on towers a hundred or more feet in height," this was a fairly tall order.

"If we were ever to entertain the slightest hope of substituting pounds for tons, and tens of watts for kilowatts," Wattson-Watt later recalled, his team would have to secure a steady supply of some highly superior form of insulation. The ideal insulating material would have to combine high dielectric strength—permitting passage of electricity without conducting current—with low weight. Fortunately, some smart chaps at the Imperial Chemical Industries Research Laboratories in Winnington had recently come up with the right stuff.

Polyethylene

Well, you should see Polythene Pam.
She's so good-looking but she looks like a man.
She's the kind of a girl that makes the news of the world.
Yes, you could say she was attractively built.
                              —JOHN LENNON AND PAUL MCCARTNEY

In early 1933, E. W. Fawcett and R. O. Gibson, two young organic chemists at I.C.I.'s Winnington Research Laboratories, embarked on a long-term study to observe the behavior of various chemicals subjected to high pressures—up to twenty thousand atmospheres. Because they happened to have a loose cylinder of ethylene lying around, and were only too eager to get started, the first reaction they set off was between ethylene and benzaldehyde, under two thousand atmospheres of internal pressure.

On Friday, March 24, 1933, Fawcett and Gibson pumped twenty cubic centimeters of their benzaldehyde-ethylene gas mixture into a high-pressure reaction vessel recently constructed at Winnington Labs, affectionately known as "the Bomb." After preparing the experiment and turning the pressure up, the two chemists took off for the weekend.

On Monday morning, after checking the Bomb, they were shocked and dismayed to find no pressure left in the cylinder. The Bomb, they surmised, must have developed a leak over the weekend. Eager to find the leak, the two chemists spent the rest of the day taking the apparatus apart. While dismantling a piece of steel tubing through which gas was pumped into the reaction space, Fawcett let out a low whistle: The tube looked as if had been dipped into a molten mass of white wax. After poking and prodding at the solid white glob at the tube end, Gibson paused briefly to jot in his notebook: "Waxy solid found in reaction tube."

Intrigued by the possibility that they might have produced a new plastic, Fawcett and Gibson decided to repeat the experiment—this time without the leak. On their second go-around, they again found a trace amount of white polymer inside the vessel. But on their third attempt, the Bomb blew up. Not, fortunately, with sufficient strength to cause any loss of life, but strong enough to result in "smashed jointing, tubing, and gauges, as well as the destruction of parts of the laboratory," as a damage-assessment report summarized.

Studying chemical reactions under extreme environmental conditions was, of course, the primary thrust of the research conducted at Winnington. But blowing up laboratories was another matter. The explosion was bad luck for Fawcett and Gibson, because managers the world over in 1933 were looking for excuses to slash funds for projects lacking short-term profit potential.

With the depression showing no signs of abating, despite Fawcett's strong hunch "that the new polymer might turn out to be important," Winnington research manager Dr. H. E. Cocksedge banned high-pressure studies until stronger reaction vessels could be developed. With funds lacking to devise any such vessels, Fawcett and Gibson were strongly urged to find work in other less high-pressure areas.

Fawcett took off for America to work on Wallace Carothers's team at Du Pont. Gibson remained at Winnington. But not until Dr. Michael Perrin, their senior supervisor and original proposer of the derailed high-pressure studies, decided to risk his job (and his neck) by reproducing his protégés' experiment in secret was the white waxy polymer given a second chance.

With Cocksedge's ban still in effect, Perrin elected to pursue this sensitive project in his spare time, and after hours, when most of the other lab workers had long since gone home. Six days before Christmas, at six-thirty on the evening of December 19, 1935, Perrin and a group of assistants pumped twenty cubic centimeters of commercial ethylene (guaranteed 96 percent pure) into the Bomb. While furtively glancing this way and that in a farcical attempt to ensure that the coast was clear, at a discreet signal from research engineer E. D. Manning, Perrin cranked the pressure up to two thousand atmospheres, turned the temperature up to 170 degrees centigrade, and anxiously awaited further developments.

To Manning's dismay, the pressure-gauge pointer dropped nearly to zero, indicating a sharp drop in internal pressure. While Manning ascribed the drop to a leak, Perrin was convinced that it must have been caused by the depressurizing effects of the polymerization process. After the Bomb was cooled and dismantled, Perrin opened the reaction vessel. The team was rewarded with a brief flurry of plastic snow. "You can imagine our excitement," Manning told an audience years later, "when masses of snowy white polymer came spilling out." Perrin was excited as well, but his enthusiasm was dampened by the disappointingly small amount of polymer generated—only eight grams. Not nearly enough, in his opinion, to warrant the sharp drop in pressure. As it turned out, Manning had been right all along: The apparatus had sprung a leak. But Perrin turned out to have been right as well, because the loss of pressure inside the Bomb had also been caused by polymerization.

Both turned out to have been lucky. Only after months of intensive research did Perrin and his crew of guerrilla scientists confirm that if the Bomb had not sprung a leak, and if the ethylene they had used had been 100 percent instead of merely 96 percent pure, they would have been unlikely to survive their experiment. Without a trace of oxygen left in the ethylene to act as a catalyst, and without a leak in the reaction vessel to lower the pressure, the resulting explosion would almost certainly have blown them—and the whole lab—to kingdom come.

Perrin and his colleagues had not, of course, discovered polyethylene.

They had just rediscovered it. It took I.C.I. another two years to perfect a large-enough-volume compressor to generate the sort of pressures required to manufacture polyethylene on a commercial scale. But by February 1936, I.C.I. was ready to file for its first patent on the new plastic.

Just about every plastics pioneer since Alexander Parkes had tried and failed to develop a synthetic substance superior to rubber and gutta-percha as an insulator for submarine telephone and telegraph cables. Polyethylene, as it turned out, was no better than rubber and gutta-percha for insulating ordinary high-voltage wires. But for coating the new high-frequency, multichannel coaxial cable the British Submarine Cable Company was planning to lay beneath the Atlantic to America, it was ideal. In 1938, a mile of coaxial cable sheathed in polyethylene was laid between the British mainland and the Isle of Wight. A new era of high-speed, multiplex telecommunications had just begun, rendered feasible at last by polyethylene's extraordinary dielectric properties.

These were easily demonstrated to skeptical observers by placing a thin sheet of the new plastic between two electrodes. The voltage passing between could be cranked up to maximum levels without the sheet exhibiting any visible deterioration. But if that same sheet of polyethylene were held up to a match, it would melt away like an ice cube in August.

By February 1940, polyethylene permitted Wattson-Watt to reduce his multiton radar station to a six-hundred-pound airborne gorilla known as the Mark II, soon to be supplanted by the smaller, lighter Mark III and Mark IV models.

Though the addition of airborne radar slowed a British Blenheim down to near sitting-duck status, it gave British fighters and bombers the means to detect, intercept, and shoot down German bombers at night and during foul-weather attack. Without airborne radar, the Royal Air Force would have been forced to scramble their squadrons at the sight of any approaching aircraft, vastly reducing their defensive efficiency. The capacity to mount radar on ships at sea was the decisive factor, many military analysts agreed, in the strategically critical victory of the British over the Germans at the Battle of Matapan, off Greece, and in the tracking and sinking of the Nazi dreadnought *Starnhorst*. Though the Germans eventually developed radar, they never did develop polyethylene. For the remainder of the conflict, their airplanes and ships were at a distinct disadvantage in tracking enemy attackers at sea and in the air—particularly through rain and cloud.

Plastic War

The impact of the rapidly developing defense program on our economic system now makes it imperative that certain vitally essential metals be conserved for primary defense purposes. ... This means that the whole question of plastics now becomes more important than ever before.

—E. R. Stettinius Jr.
Director of Priorities
Office of Production Management
April 1942

In May 1942, Major E. L. Hobson, a staff officer at the U.S. Army Quartermaster (QMC) Headquarters in Washington, nearly wasted a morning placing urgent long-distance calls around the country on what certainly seemed like a fool's errand. For two centuries, the regulation U.S. Army bugle, which weighed twenty ounces, had been fashioned by skilled brass founders from the finest virgin brass—three-quarters of which is copper. But with copper ranking high on the list of critical metals (for which it was a Quartermaster Corps officer's solemn duty to find substitutes whenever feasible) and with an estimated two pounds of copper consumed for every bugle made (large amounts of brass were wasted in machining), it had become Hobson's task to find someone to make—much as the thought pained him—a plastic bugle.

The U.S. Army Quartermaster Corps was imbued with a centuries-old tradition of procuring the world's finest brass bugles, leather harnesses and saddles, canvas tents, cotton and wool uniforms—every material item needed to pursue the ancient and honorable art and craft of war. But as Napoléon once said, "Wars are won or lost by the little things." And as Lord Plumer, one of Britain's great World War I quartermaster generals, had more recently put it: "Strategy is subservient to supply." Though Major Hobson was as emotionally attached to the beauty of brass as the next QM, if the brassless bugle was able to meet tough army specs, he didn't care if it were made of Swiss cheese.

Before breaking for lunch and abandoning this crazy scheme, Hobson decided to place one last call, to Elmer Mills of Mills Plastic Products in Chicago. Mills had earned a can-do reputation in the plastics trade by filling a multimillion-unit order of red plastic trumpets for the previous Christmas season.

Now, Major Hobson appealed to Elmer Mills to heed the call of his country.

"No problem," replied Mills.

Now it was Hobson's turn to be blunt.

"Any bugle you make for the United States Army is going to have to meet some extremely stiff performance requirements."

There was an extended silence at the other end, explained by the fact that Mills didn't like people reflexively disparaging plastic, particularly if they didn't know the first thing about it. "If you want a plastic bugle," Mills retorted curtly, "we can make one for you."

Two weeks later, Mills called Hobson back.

"We've got your bugle."

"Wonderful," Hobson replied. "Why don't you shoot it down to me for testing. If it works out, we'll send for you in a couple of weeks and talk it over."

"Sorry, but this is the only plastic bugle on earth. I'm not prepared to ship it anywhere. If you're interested in taking a look at it, I'll be in your office at nine o'clock in the morning the day after tomorrow. Have your people ready to test it. If it checks out, we'll consider making a few more for you."

What Mills neglected to tell Hobson was that he had spent the last two weeks holed up with Mr. M. H. Berlin of the Chicago Musical Instrument Company, for whom he had molded those two million toy trumpets. Intrigued by the idea of obtaining a hefty government contract, and by the satisfaction of helping his country out of a jam, Berlin had called in Frank Aman, who just happened to be the leading designer of woodwind instruments in Chicago, if not the country. A Hungarian immigrant and proud scion of a long line of musical-instrument makers, at eleven Aman had apprenticed to his uncle Irwin, the finest woodwind maker in Hungary. After putting in a few years serving as plant foreman at the world-famous Hohner woodwind factory in Hamburg, Aman had emigrated to Chicago, where he set up shop as a maker of fine flutes, clarinets, and other wind instruments for concert musicians.

A man of Aman's reputation and devotion to musical tradition was not, at first go-around, enthusiastic about designing a plastic bugle. But after learning that this was the U.S. Army Quartermaster Corps calling, and that this concerned the urgent need to save brass, Aman dropped everything to delve into two weeks of around-the-clock work with Berlin and Mills.

When Mills and Berlin arrived at Hobson's office at the Pentagon, they were surprised to find a very tall soldier wearing a sergeant's insignia standing at ease beside Hobson's desk. Hobson introduced his guests to the man widely reputed to be the best bugler in Washington—the solo cornetist of the U.S. Army Band. When the tall soldier bent stiffly at the waist to pick up the plastic bugle, he couldn't help grimacing. Weighing in at just under ten ounces, the bantamweight bugle felt like one of the plastic toys Elmer Mills had molded for the Christmas season.

Following Hobson's orders, the cornetist played taps, before moving onto reveille. Then on to "Boots and Saddles." At a quick nod from Hobson, the party adjourned to the Washington Mall for an open-air test, for which Hobson had recruited the solo cornetist's commanding officer, Captain Darcy. Obligingly positioning himself half a mile away in order to judge the sonic carrying power of a plastic versus brass bugle under strict field conditions, Captain Darcy surprised himself by—sight unseen of course—giving the plastic the nod.

For round three, a trio of solemn-faced gentlemen from the Army Quartermaster Technical Committee commenced proceedings by hurling the bugle against a brick wall. After picking it up off the ground, three acoustic technicians donned headphones to subject the rebounded bugle to stroboscopic tests of the frequency of each note. Not only had it not broken or dented with the force of impact, but its tonal quality had not been measurably affected. At 4:30 that afternoon, Mills and the Chicago Musical Instrument Company were given their order for two hundred thousand plastic bugles—the fastest new-product evaluation in QMC history.

Before you could say "Jack Robinson," Major Hobson's hastily formed "Plastics Sub-Unit of the Quartermaster Corps, Standardization Branch," officially charged with "introducing plastics into Army equipment wherever and whenever it represented a significant improvement, economy, or conservation of critical materials," were hot on the trail of Mother Nature. Their mission: to root out strategically critical natural materials ripe for replacement.

Brass was to be conserved wherever practically feasible, which put next on plastic's hit list the legendary brass sergeant's signaling whistle, respectfully known throughout the armed forces as "the Thunderer." "The assembled plastic signal unit replacing it," maintained one QMC report, "not only looks as businesslike as the brass original, but hangs less heavily on the whistler's lip."

Slated next for supplanting was the traditional steel M-1 high-pressure helmet liner, which according to one press report had to serve a soldier as "more than just a tin hat. It is a battle turret, a shock absorber, and in a pinch, a wash tub." Hobson's choice was phenolic resin, required by stiff QMC specs to be "dull, scuff-and-chip proof, resistant to sun, frost and the chemical stream of the delousing process." Delousing alone required the liner to withstand temperatures in excess of 250 degrees Fahrenheit for half an hour.

It had to withstand temperatures from 40 degrees below zero to 160 above. It had to resist a twenty-pound blow to the head. And since it had to serve as a sun helmet in the tropics, it could not under even the most extreme climactic conditions delaminate. It also had to withstand the impact of a .45-caliber pistol.

At a conference in Washington convened to mobilize the plastics industry, Lt. Edward T. McBride, a technical officer with the army's Office of the Chief of Ordnance, laid down the laws of plastic survival in wartime. Of the seven hundred plastic firms present, a third would be out of business within a year, McBride predicted, if they didn't wrangle defense contracts.

"Cut out the mystery, glamour and charm," advised Jesse Lundsford, a principal engineer in the Research Branch, Design Division, Standards and Tests Section of the navy's Bureau of Ships. "That was good sales talk when you had plastics to sell, and few wanted to buy. But for the war services, just let plastics grow up and mature—their service performance can keep them rolling from now on. If you have a plastic that's new, why not say so, simply and without glamour, tear-shedding, flag-waving or voice-quaver. I assure you, we are listening."

Proof of the pudding was not forthcoming until July 1942, when in response to the occupation of Norway by German forces, the navy QMC announced with grave concern that the traditional navy dress sword-handle grip, which for centuries had been wrapped in the skin of a baby Norwegian shark (bleached white to match the dress uniform), would henceforth be covered with injection-molded cellulose acetate. "To retain the mottled effect of the pebble-grained surface, the new grip will be wrapped barber-pole fashion with a gold-colored brass wire," the communiqué soberly advised. Though still too scarce for an army bugle, it was still possible for the Pentagon to scare up a little brass to dress up a ceremonial dress sword.

Offensive Nylon

"Everyone is well aware of the fact that nylon has gone to war," announced a Du Pont spokesman, D. D. Payne, in March 1942, after the government requisitioned all nylon production to be channeled into defense. Critical needs for nylon seemed to multiply by the day, from glider towropes to parachutes to cords for synthetic rubber tires, which heated up faster and ran at much higher temperatures than real rubber. "Nylon for stockings," Payne solemnly intoned, "has become nylon for parachutes, and dozens of other defense-related purposes," which, since loose lips sank ships, he was not at liberty to divulge at the present time.

The requisitioning of nylon had dealt a serious blow to a hosiery industry still reeling from the earlier requisition of silk. Which left the women of America, and the hosiery mill owners, with little to work with but rayon (which shrank and distorted if washed in water) and high-tricot cotton "lisle."

In a desperate bind, the hosiery industry urged stores to give "increased promotional stress to hosiery in order to prevent a possible bare-legged fad before it has a chance to start." But the watchword of the industry, "Keep women stocking-minded!" quickly became a poignant rallying cry for a lost cause. Despite glamour-girl Betty "Legs" Grable's well-publicized auctioning off of one pair of her nylon stockings for $40,000 in war bonds, a patriotic "bare-legged look" was catching on, which proved nearly as alluring as Rosie the Riveter in her trademark slacks. Not long after Pearl Harbor, a national convention of women's clubs sent a letter to the Office of Production Management (OPM) offering to pass a resolution "banning hose of all kinds if it would help the war effort."

Nylon stockings could still be had—for a price—on a thriving black market. In 1942, thirteen cases of raw nylon en route to a parachute plant in Winston-Salem, North Carolina, were mysteriously diverted at a motor-freight terminal in Greensboro, North Carolina. An idle hosiery mill, accepting the inane cover story that the nylon had been salvaged from a "warehouse fire," made it up into hosiery, stretching the precious contraband as far as possible by making the tops and toes out of cotton. The stockings sold for $5 a pair to bootleggers, who retailed them on the street for $10. Nylon worth $7,800 at the factory was now worth, on the street, nearly $150,000.

A few months later, the five thousand pairs of contraband stockings were confiscated at the mill by the FBI, who proceeded to arrest a former official with the trucking company and two hosiery mill men, all three of whom ended up doing serious time. By court order, the five thousand pairs of nylons were sold at the U.S. marshal's office in Greensboro for a government-ceiling price of $1.65 a pair. The sale was slated to begin at ten in the morning. By five the night before, the lines began to form, and when the doors opened, a line of women standing four abreast stretched down four city blocks. Half of them went away empty-handed.

American women not content with the bare-leg look, or sick of wearing saggy, baggy rayons, grew so desperate for the real thing that they fell for con artists who brazenly peddled spurious compounds that, after being dissolved in water, were guaranteed to "nylonize" rayon stockings. After word of that scheme got around, one major hosiery maker wryly commented, "If any chemist has a formula for that, he needn't bother with stockings. I'll give him $5 million in cash."

Still available on a piecemeal basis for strategic purposes, a pair of nylons possessed the underworldly allure of an adventure with Mata Hari. Rather than being "sent to Iceland on lend-lease," as one irate congressman announced on the floor of the House, thousands of nylon stockings were supplied to Allied secret agents, who found that they were worth more than their weight in gold when supplied as bribes to corrupt officials. According to one hush-hush report, "Half a dozen pairs of nylon stockings will buy more information from certain mysterious women in Europe and North Africa than a fistful of money."

In preparation for the Allied landings in North Africa in the summer of 1942, a top-secret order for nylons went out from the War Department to a few hosiery mills. Three hundred pairs of nylon stockings were shipped off to the headquarters of Gen. Mark Clark, commander in charge of the North African landings, "for immediate distribution to key French women and certain French officials as a first step in the softening-up process preceding the landings." Other women, not so fortunate, were forced to resort to drawing lines on the backs of their legs with eyebrow pencil to simulate seams.

I Don't Care If It Rains or Freezes, Long As I Got My Plastic Jesus

In late 1941, Cardinal Hinsley, the right honorable archbishop of Westminster, announced the imminent distribution of two and half million plastic crucifixes to all members of the British fighting forces, upon request. Molded from white phenolic resin and measuring 2 by 2½ inches, the crucifixes came equipped with a tiny ring on the top so that they could be worn around the neck. The figure of Our Lord had been sunk into the cross to keep the cross from catching on the wearer's uniform.

The Vinyl Solution

> Without vinyl, our troops might have been forced to forgo such comforts as waterproof tents, groundsheets and raincoats, ponchos—no small comfort to one who has to struggle with wet tent flaps and sleep in wet uniforms.
>
> —DU PONT OFFICIAL JAMES OWENS

In 1909, the eleven-year-old Waldo Lonsbury Semon was too busy carrying buckets of water to the thirsty elephants who annually invaded his hometown of Medford, Oregon, as the advanced detail of Buffalo Bill Cody's Wild West Show, to pay much attention to Leo Baekeland's invention of Bakelite. But seventeen years later, in 1926, organic chemist Waldo Lonsbury Semon was assigned by his employer, the rubber company BFGoodrich, to develop a new synthetic material for binding rubber linings to metal tanks.

Two years before, E. W. Reid of the Carbide and Carbon Chemical Corporation had patented a copolymer of vinyl chloride and vinyl acetate. At around the same time, two I.G. Farben chemists had patented a nearly identical compound. But the problem with all the vinyls produced to date was that they tended to break down and lose resiliency when processed at high temperatures. Waldo Semon's key contribution was devising a way to make vinyl industrially processable by mixing a high-boiling-point liquid (tritotyl phosphate) into the vinyl chloride, thereby ensuring that it would remain stable in high heat.

Semon's seminal substance, formally known as polyvinyl chloride, or PVC, was practically stillborn, since industrial uses for this new water-resistant, fireproof material—which could be molded or extruded into sheets or film—were not immediately apparent. Charged by Goodrich with finding some practical use for the stuff or face a cutoff of funds for future research, Semon experienced a commercial epiphany while watching his wife, Marjorie, sew a set of shower curtains out of rubber-lined cotton.

Vinyl, it struck him in a flash, would make the perfect waterproof fabric, from shower curtains to raincoats. Not since Charles Macintosh invented the Mack had any new compound come along that compared favorably to rubber when it came to foul-weather gear. Conventional rubber storm gear tended to be heavy and exude an unpleasant odor when wet. But vinyl beat rubber hands down. "From the teeming torrents of the untamed tropics, to the bite of the arctic snows, and the damp, congealing London fog, ruthless Nature provides a new setting for every theater of war," as one vinyl advertisement breathlessly put it.

At sea, vinyl's resistance to fire and flame—a sailor's worst nightmare—made vinyl ideal for ship's upholstery. Its "variety and brilliance of color" made it the obvious insulation of choice for "color-coding electrical wires to designate different circuits on landing craft." Soldiers packed into watersoaked landing craft learned to appreciate their new waterproof vinyl "gun boots," which kept their guns from getting wet and from sinking beneath the waves if they fell overboard.

Saran

An overturned ink bottle streaking the spanking newness of the living room sofa; a blob of grease sullying the automobile upholstery; unsightly rain and dust spots defacing the beauty of brightly hued draperies—each of these accidents spells one of life's darker moments for the average householder. But not for long . . .

—DOW CHEMICAL AD, AUGUST 1942

Yet another vinyl-based compound, Saran, had been originally envisioned by its manufacturer, Dow Chemical, as a surefire covering for theater seats. "Unlike ordinary fabrics, which almost never yield chewing gum once it has taken hold," Dow maintained, "virtually any commercial cleaning fluid will dislodge from Saran this bane of the movie-theater manager's existence." Another prewar-satisfied Saran customer was the New York Transit Authority, which ordered thousands of yards of open-weave-mesh monofilament fabric for replacing the badly soiled woven-rattan seats on its subway cars.

Saran was certainly chewing-gum-proof, but it was also impervious to rot. Which made it an ideal candidate for nonrotting woven mesh for insect screening in the tropics. In 1942, George Knox, chief of the navy's Bureau of Yard and Docks section, asked Jack Holman, head of the Chicopee Manufacturing Company (producers of surgical gauze and tobacco cloth), to run up a prototype Saran-mesh tent. It stood for months between two wings of the Navy Department building in Washington while navy officials gave it the once-over. Its sheer sides earned it a coy nickname: "the Waves' dressing room."

Defense contractors soon discovered other, even more exotic uses for Saran: When sprayed in film form on planes being shipped overseas, fighters were so well protected from salt and sea spray that they could ride tied down on the flight deck of an aircraft carrier, in the open air. They no longer needed to be disassembled, packed in grease in wooden crates, laboriously degreased at their destination, and painstakingly reassembled before being put into action.

At this moment, men on our fighting fronts throughout the world are gaining first-hand knowledge of a new transparent packaging material. When they unpack a machine gun they find it protected from moisture with *Saran Film*. There are no coatings of grease to be removed—no time lost. The gun slips out of its *Saran Film* envelope clean, uncorroded—ready for action!

This Dow product is undoubtedly the packaging development of the year. Imagine a transparent film—tough, flexible and not only strongly resistant to chemicals, but actually possessing three times more moisture resistance than any other comparable material. These are some of the many superior qualities of *Saran Film*. Such a combination of advantages has far reaching implications that suggest wide usefulness in a peacetime world.

Action Scene Above from Buna Campaign. —Official photo by U. S. Army Signal Corps

Delivered in Saran Film
Ready for Action

THE DOW CHEMICAL COMPANY
MIDLAND, MICHIGAN
New York · St. Louis · Chicago · Houston
San Francisco · Los Angeles · Seattle

Saran film
KEEPS MOISTURE IN ITS PLACE

DOW
CHEMICALS INDISPENSABLE TO INDUSTRY AND VICTORY

Courtesy of Dow Chemical

"Not only planes, but also artillery and other sensitive military equipment were being shipped to far-flung battle fronts adequately protected from an ever-present enemy," crowed a patriotic Dow Chemical ad. "They are encased in a new revolutionary transparent packaging material . . . which provides three times greater moisture resistance than any other comparable material." Watch out, cellophane.

Crystal Ball

The B-19 Douglas Superbomber, known as the "Guardian of the Hemisphere," first rolled off the line in August 1941, with a flight range of 7,750 miles—enough to fly from Los Angeles to London and back to New York without refueling. Its wingspan of 212 feet was more than twice that of any commercial airliner. Its fuel capacity of 11,000 gallons was equal to that of a railway tank car. Its normal gross weight, 140,000 pounds, was comparable to that of a shipping vessel.

But its most revolutionary feature was the bombardier compartment, which was fully protected by a transparent acrylic plastic sheet bent by heat into a blister resembling a gigantic crystal ball. The nose turret was likewise enclosed in a heat-formed transparent plastic bubble. The same went for the gunner's compartment, likewise fabricated from transparent methyl methacrylate sheets. Plexiglas, the material GM used to awe all comers with its transparent plastic Pontiac (and helped RCA make waves with its transparent TV) had since its debut at the 1939 New York World's Fair been brilliantly adapted to modern warfare. The "yen for transparency" that had seized the American imagination in the years leading up to the war had since become a deadly serious business.

Plexiglas might never have been invented if, on a seasonably warm summer day in August 1904, Dr. Otto Rohm, an organic chemist employed by the Municipal Gas Works in Stuttgart, Germany, had not become violently ill from a wave of sickly-sweet odors that came wafting into his office from the local tannery, borne on a sticky summer breeze.

The obvious culprit, Rohm was all too aware, was the revolting "bating" process by which dehaired hides were chemically softened with fermented dog dung before they could be tanned. After cleaning himself up, it occurred to Rohm that the tannery smell was distinctly reminiscent of a gaseous waste produced by his own works—which was one of the reasons the municipal authorities had chosen to locate the two facilities so close together.

The "gas water" produced by Rohm's gasworks was a repugnant combination of ammonia, carbon dioxide, and hydrogen sulfide. Since it smelled so much like dog dung, Rohm wondered whether it might be possible to synthesize a superior bating solution from coal-tar derivatives and molecules drawn from the air, thereby eliminating the troublesome canine factor.

After successfully "bating" a batch of goatskins with a synthetic compound derived from coal-tar distillates, Rohm resigned from his

Courtesy Rohm & Haas Company

position at the gasworks and wrote to his old friend Otto Haas, a recent émigré to America, to look into forming a partnership to market Oropon—German for "juice"—worldwide.

In his letter to Haas, who had settled in Philadelphia, Rohm asked him to investigate the American tanning industry's annual consumption of dog dung. Haas was receptive to the idea of a joint venture, and had soon joined forces with Rohm to form Rohm & Haas, with American offices in Philadelphia and German offices at Darmstadt, the site of Rohm's lab.

By 1920, the company had grown sufficiently prosperous on its synthetic dog dung to branch out into other, potentially more lucrative, areas. Rohm was now free to return to his first love, acrylic acid chemistry, a subject on which he had written his 1901 doctoral dissertation at the University of Tübingen. After eight years of intensive research into the behavior of acrylic esters and polymers, Rohm proudly introduced the first commercial acrylic product: a methyl methacrylate interlayer for automotive safety glass called Plexigum.

Unbreakable or flexible glass had been the object of chemical quests since before the days of the alchemists, and a popular subject of legend and folklore. During the reign of the Roman emperor Tiberius, a certain glassblower (so Petronius reports) produced a transparent vessel that could allegedly be bounced on a stone payment, like a ball, without breaking.

After being granted an audience with the emperor to demonstrate his miraculous invention, the proud inventor was promptly dispatched to the palace executioner. His capital crime: displaying the arrogance to invent a product so valuable that it threatened to devalue the gold hoard in the imperial treasury. Though rumored to have been possessed by the ancient Egyptians, flexible glass didn't resurface as a commodity until 1881, when the German chemist Kahlbaum (1853–1905) astonished a group of companions in a Swiss *Bierstube* (beerhall) by displaying an unbreakable beer glass, no doubt fashioned from the methyl acrylate polymer he had recently synthesized. But Kahlbaum, a pure scientist, never attempted to commercialize his discovery, thus leaving the field open to the Parisian poet, painter, philosopher, solicitor, musician, and scientist Edouard Benedictus (1879–1930) a dreamy romantic who claimed descent from the medieval Jewish philosopher Baruch Benedict de Spinoza. One day in 1903, as the dilettante Benedictus would later somewhat hazily relate:

> I was setting my laboratory to rights, moving the glassware around, when a bottle slipped from my hands and fell to the floor from a considerable height. This bottle, of about one liter size, I picked up apparently unbroken. The glass was starred like Bohemian crystal but was held firmly together by some internal adhesion.
>
> The bottle had contained a solution of nitrocellulose, from which, over a period of fifteen years, all the solvents had evaporated, thus lining the interior of the bottle with a celluloid coating of great strength.
>
> So firmly were the fragments of glass held by this layer that not a single piece of any size became detached or even loosened. Having made a thorough examination of all parts of the bottle, I attached to it a label reading: "November 1903—this flask fell from a height of 3.5 meters, and was picked up in its present condition." I replaced the bottle on its shelf and

thought no more about it until one day when my attention was attracted by two accidents caused by the breaking of glass in vehicles. Two young women suffered lacerations of the neck in the collisions.

I was sitting after dinner thinking intently of these two accidents when suddenly, without warning, a faintly illuminated image of my bottle, moving as though alive, appeared upon the wall. Emerging from my reverie I rose to my feet, entered my laboratory, and gave myself up to the deliberate contemplation of the possibilities of this embryo of an idea which had come to me from my bottle. At dawn the next morning I found myself still with the bottle in my hand, not having stirred for nine hours, but with a program of experimentation drawn up, which I proceeded to carry out step by step.

Benedictus may have been a dreamer, but he knew enough science to file a publishable patent. In 1909, he founded La Société du Verre Triplex in France, and a British subsidiary, the Triplex Safety Glass Company, to market celluloid-laminated glass to the infant automobile industry. But with average speeds still too low to result in many life-threatening accidents, Benedictus's glass was mainly used in racing cars—until 1928, when at Henry Ford's insistence the Model A became the first car to come equipped with a standard-issue safety-glass windshield made from a celluloid interlayer. Unfortunately, those early laminated windshields tended to yellow and grow hazy with prolonged exposure to sunlight.

In 1931, a triumphant Otto Rohm produced a clear sample of polymethyl methacrylate sheet that wouldn't soften or bend in the heat until warmed above 110 degrees centigrade. It could be easily worked and shaped with standard saws and drills. Rohm was so confident that he had hit the jackpot with the first commercially viable safety glass that he sold off production and development rights to I.G. Farben in exchange for funding future research.

With his chief assistant, Dr. Walter Bauer, Rohm fashioned a sandwich from two plate-glass panes and poured methyl methacrylate monomer into it. When the sandwich cooled, they carefully pulled the two panes of glass apart and were pleased to observe that the plastic sheet, rather than sticking to the glass, cleaved cleanly into a smooth, transparent sheet.

Rohm was so enthralled by his "organic glass" that he insisted upon hatching new applications for it by the day. He proudly sported the world's first pair of acrylic eyeglass lenses, and replaced the side windows (successfully) and the windshield (unsuccessfully) on his Mercedes Benz touring car with cast acrylic sheet. Acrylic sheet's one deficiency when stacked up against glass was a tendency to scratch easily.

Rohm's former partner, Otto Haas—the two companies had long since parted ways—retained U.S. rights to all Rohm's inventions in exchange for funds to underwrite acrylics research. To enhance his own production capacity for methyl methacrylate sheet, Haas sent Dr. Donald Frederick, his favorite staff chemist, to Rohm's lab in Darmstadt to learn the fine art of Plexiglas manufacturing.

In January 1935, Frederick spent two eye-popping months observing all phases of Rohm's operation, except for the forming of flat sheet into three-dimensional shapes, which the Nazis did not want Frederick to observe, since it involved outfitting the cockpits of Luftwaffe bombers with Plexiglas hoods, in violation of the Versailles accords.

After returning to Philadelphia, Frederick took a few samples of Rohm's acrylic sheet down to Washington to show to technical officials of the army air corps. As a means of promoting his new product, Frederick persuaded General Motors to let him work with Fisher Body to produce the famous Plexiglas Pontiac that wowed the crowds in the rotunda of the GM Futurama.

As war threatened to break out in Europe, the U.S., British, French, and German air forces all raced to replace conventional glass in fighter-plane cockpits and bomber enclosures with Plexiglas. As the war heated up, Frederick gained expertise in blow-molding and vacuforming these transparent blisters into spherical shapes. Not only did these require far fewer workers to install than had been needed to cast and fit flat acrylic sheet, but Frederick's innovative bubbles and blisters increased the effectiveness of the aircraft as a weapon by giving pilots and gunners unbroken panoramic views of the hostile skies.

In 1942, the John Wesley Hyatt Award was duly presented to Dr. Donald S. Frederick of Rohm & Haas "for his work adapting Plexiglas to bomber noses and other large-size plastic sections, which aviation experts agree, give our planes an advantage not held by the enemy."

Acrylic polymers, as it turned out, possessed strategic potential beyond their ability to be molded into transparent plastics. One of Rohm & Haas's leading chemical researchers, Dr. Herman Bruson, a native of Middletown, Ohio, a graduate of MIT and of the Polytechnique Institute of Zurich, had spent years before the war investigating the esters formed by the reaction of acids with the so-called higher alcohols.

On one occasion, Bruson dissolved a certain ester in mineral oil. After testing the oil, Bruson found that the polymer additive had flattened the oil's viscosity index without appreciably thickening the oil itself. Typically, oil gunks up in the cold, clogging the valves and pistons of internal combustion engines called upon to operate at extremely low temperatures. But when this methyl methacrylate polymer—a close relative of Plexiglas—was added to oil, the oil continued to flow freely even at subzero temperatures.

Rohm & Haas never saw fit to commercialize Bruson's discovery. But in early 1942, a member of the National Research Defense Committee—a civilian arm of the War Department—began combing through old patent files for unexploited inventions that might possibly aid the war effort. After stumbling across Bruson's patent and instantly grasping its strategic implications, he dispatched an urgent wire to Philadelphia asking Bruson to concoct a test batch to be sent to Washington.

After putting it through its paces, the technical department of the army air corps was delighted to report that Bruson's additive (Rohm & Haas called it Acriloid) kept airplane hydraulic fluid fluid and functional at ambient temperatures below minus 150 degrees Fahrenheit.

In the spring of 1943, a Russian general celebrating the Soviet victory at Stalingrad (which decisively turned the war on the eastern front against Hitler), wrote jubilantly to Herman Bruson, care of Rohm & Haas, congratulating him for his decisive role in winning the Battle of Stalingrad. While the Nazi tanks and artillery equipped with conventional oils and hydraulic fluids ground to a halt in the bitter cold, Soviet tanks and artillery ranged up against them had functioned perfectly, no matter what the temperature. As Napoléon said—and he should have known—wars are won and lost on the little things. In the deadly chill of a long hard Russian winter, a few gallons of methyl methacrylate ester sufficed to turn the tide.

Teflon

In September 1942 Lt. Gen. Leslie Groves, commanding general of the Manhattan Engineering District, slipped into Wilmington to brief a select group of Du Pont executives on the nature and scope of his top-secret assignment. He was seeking their help in designing and constructing a massive plutonium separation plant to be located in the desert area of eastern Washington State, to be known as the Hanford Engineer Works.

For the impressive sum of one dollar—Du Pont was willing to give its planning expertise away in exchange for the promise of fuller involvement down the road—Du Pont drew up blueprints for the vast complex of reactors and support buildings slated to rise at yet another of the Manhattan Project's mysterious and remote "secret cities," this one in the wilds of Washington. Du Pont even created a new division to carry out the classified work, which it code-named "TNX."

One of the toughest problems confronted by the nuclear physicists studying the strategic aspects of fission at the top-secret government laboratory at Oak Ridge, Tennessee, was separating the 1 percent of fissionable isotope uranium 235 from the uranium 238 in natural uranium.

In employing a method known as "gaseous diffusion," the physicists required a form of fluorine gas, which bears the distinction of being one of the most corrosive substances on earth. Asked to come up with some sort of miracle material capable of withstanding this harsh gas's corrosive effects, Du Pont's TNX immediately thought of PTFE, a remarkable new compound discovered by accident four years earlier by a green chemist still in his twenties.

As a protégé at Ohio State of the future Nobel Prize winner Paul Flory, Roy Plunkett had specialized in fluorine chemistry, which formed the basis of refrigeration research. Du Pont's Jackson Laboratories, a leading center in the field, hired Plunkett straight out of graduate school to conduct a course of experiments on the behavior and properties of a particular fluorohydrocarbon gas known as Freon, the most commonly used refrigerant.

As a first step in his experimental regime, Plunkett prepared a hundred pounds of tetrafluoroethylene gas, pumped it into several small cylinders, sealed the valves, and left them sitting overnight in cold storage. The next morning (April 6, 1938) accompanied by his assistant, Jack Rebok, Plunkett fit one of the gas cylinders to his lab reactor, opened the valve, and waited for the gas to whistle into the chamber.

Nothing happened. No gas appeared to be escaping from the cylinder. Checking the pressure gauge, Rebok and Plunkett were shocked to find the needle pointing at zero.

"Hey, Doc," Rebok asked Plunkett, scratching his head. "Did you use all this stuff up last night?"

"I don't think so," Plunkett replied, baffled.

"Nothing's coming out," Rebok muttered, tinkering with what he assumed to be a broken valve.

His curiosity aroused, Plunkett placed the errant cylinder on a scale. Sure enough, it weighed just the same as it had the night before, which meant that something was in there. But what? Removing the valve, Plunkett turned the cylinder over. A tiny trace of white powder spilled out. Snatching a long wire from his lab bench, Plunkett stuck it into the cylinder and began scraping away at the bottom. With each new shake of the cylinder, more white powder spilled out. Like a stage magician, Plunkett then took a hacksaw and sawed the cylinder in half. Coating every inch of its inside surface was more of the mysterious white solid.

More remarkable than the reaction, which had been entirely spontaneous, was the nature of the reactant itself. Like a true off-spring of Bakelite, soldering irons would not melt it, nor electric arcs char, this strange variant of polyethylene—the hydrogen atoms of the ethylene molecule replaced by atoms of fluorine. Like its distinguished ancestor, it would neither burn nor melt—not below 620 degrees Fahrenheit—or two hundred degrees higher than the melting point of tin. And then, instead of turning into a liquid, it congealed into a translucent gel. Extreme cold had no effect on it, even at temperatures just above absolute zero.

Tests revealed PTFE—polytetrafluoroethylene—to be a solid form of tetrafluoroethylene gas, which had spontaneously polymerized inside the cylinder. Teflon, as it soon would be known, was radically unlike any polymer ever conceived by man or nature. It would not conduct electricity—at all. Because it didn't combine with oxygen, and couldn't sustain life, it was immune to attack by mold, fungus, or other bacterial pests. No solvent on earth could dissolve, mar, or corrode it. Acids that easily dissolved gold and silver had no effect on it. Bathing it in boiling nitric and sulfuric acid merely gave it a good cleaning. But apart from its extraordinary durability, the most remarkable property of this virtually impregnable material was a surpassing slipperiness. It was so unaffected by the forces of friction that even chewing gum wouldn't stick to it. It was slipperier than nature's slipperiest substance: wet ice on wet ice. It would later be used—strictly for demonstration purposes—as synthetic ice in skating rinks.

Just the same, neither Plunkett nor his superiors at Jackson Labs had the faintest idea what to do with it. Not until the Oak Ridge boys urgently demanded a noncorrosive material to coat the valves and gaskets needed for separating uranium by gaseous diffusion did the inexorable imperatives of the Manhattan Project propel PTFE into action. When told that the experimental plastic would be expensive, General Groves replied that price was no object. And so, Du Pont began selling Teflon to the military in 1944, and not only to the Manhattan Project. It turned out to be a superb noncorrosive coating for artillery-shell nose cones, an ideal insulation for radar wiring (shades of polyethylene, its carbon equivalent) and a high-performance lining for storing tanks of liquid fuel so cold that conventional linings simply froze and turned brittle.

Slated to evolve into the original space-age material, Teflon would in time provide a significant boon to outer-space exploration, as well as a long-wearing lubricant for bridges that sway excessively in strong winds. But it was a reflection of the banal short-term future of the synthetic revolution that when Teflon finally made it into the hands of a grateful public, it would humbly serve on the domestic front as a convenient coating for nonstick muffin tins, skillets, and the next generation of steam irons.

Respect

R-E-S-P-E-C-T
Find out what it means to me.

—ARETHA FRANKLIN

On September 1, 1939, as Nazi warplanes swooped down over Warsaw, a tiny pilot polyethylene production plant went onstream at I.C.I. The day Japanese troops stormed into Singapore, just a few pounds of synthetic rubber had ever been produced outside of Germany. But as the war progressed, the Allies and the Axis powers found themselves locked in a synthetic struggle to replace scarce natural materials with plastics and polymers. While the giant refineries of the I.G. Farben conglomerate made vast strides in producing styrene plastics and other chemically based commodities, the nation with the greatest head start in the realm of synthetic chemistry was repeatedly outflanked by the Allied brain trusts.

From the Buna debacle to the failure to synthesize polyethylene—which the Nazis nearly acquired from I.C.I. in exchange for Buna, only to be rebuffed months before the outbreak of hostilities—the tight Nazi grip on the Farben monolith (which took over chemical plants in the occupied Eastern territories from Poland to Czechoslovakia and used slave labor to run them) in no small part contributed to Farben's failure to keep pace with Allied advances. For that failure, the Axis armies, navies, and air forces were destined to pay dearly.

Allied bombers targeted Farben factories as among the most strategically critical installations. While the sprawling BASF works in Ludwigshafen, on the battered banks of the Rhine, was blown up too many times to ever be rebuilt, on June 11, 1943, wave after wave of Allied aircraft dropped 15,660 bombs on the giant Buna plant at Huls, killing 186 workers and wounding 752. In four months, the Huls plant had been rebuilt to maximum production capacity, and continued to operate until March 1945, when Hitler dispatched a special army unit with instructions to destroy it so "that the enemy would find nothing in its place."

Dr. Paul Bauman, the plant director, persuaded the army destruction unit to disobey orders, leaving a skeleton staff in place to welcome the conquering U.S. troops. In late 1945, British troops replaced the Americans in that part of Germany. The British occupational government gave Bauman permission to reactivate the Buna plant as a means of driving the price of natural rubber, which had suspiciously soared through the roof, down to more reasonable levels.

Dr. Fritz ter Meer, Buna's chief developer, who had once memorably described natural substances as "wild horses that must be broken to reins," was soon to stand trial at Nuremberg along with Otto Ambros and nearly all the other member of Farben's *Vorstand* (board). Though Ambros, ter Meer, and many of Farben colleagues were tried and sentenced at Nuremberg for the use of slave labor and depraved indifference to human life, all Farben defendants were let off with light sentences because of a much criticized reluctance on the part of the Nuremberg judges to severely punish civilians, no matter how vile their crimes.

To those who had fought with the armed forces overseas, vinyl and Saran, nylon and Plexiglas and Lucite and the other synthetics had been trusted partners in combat, tested in climactic conditions ranging from the hottest, wettest tropical jungles to the coldest, driest Arctic peaks. No matter how unfriendly the conditions, plastic materials had been found to withstand pressures and stresses that might have frozen, melted, or rotted lesser stuff. In the tropical theater, even the cages used to keep carrier pigeons in captivity had to be made out of plastic so they wouldn't rot. The same went for jungle boots, screened tents, waterproof clothing, life rafts, canteens, and bayonet scabbards.

Anything a natural material could do, a synthetic material could apparently do better. If before the war plastic's image had been defined by frivolity, trumpery, and above all, a sleazy pretense at luxury, by the time Teflon was enlisted in the atom bomb effort, plastic materials had matured under fire. But with a vast plant capacity still geared up to produce synthetic swords, the overwhelming question became: Come peacetime, how many plastic plowshares could the public be persuaded to buy without gagging?

8

Plast-O-Rama

Courtesy of The Hagley Museum and Library

Nylo-mania

At eight o'clock on a Tuesday night in February 1946, a radio announcer for WABC in Washington, D.C., smoothly intoned a late-breaking bulletin: Any listener who called District 6363, ASAP, stood a chance of receiving a brand-new pair of nylon stockings. A local shoe store, the announcer explained, had recently received its first shipment of nylon stockings since the end of the war. The owners had elected to sell one pair each to the first thousand women who called—the fairest way they could think of to dispose of the merchandise.

The store took every reasonable precaution to prepare for the expected inundation, including the hiring of nineteen telephone clerks to staff a temporary bank of phones on the night in question. But as the evening wore on, the temporary help sat around filing their nails, waiting impatiently for the phones to start ringing. After a couple of hours of resounding silence, the store manager picked up the phone to call his boss at home for advice. To his shock, the line had gone dead.

Unbeknownst to the store, and to the radio station, the mere mention of a nylon sale had set off a telephonic tidal wave throughout the nation's capital. Within minutes, the Capital, Alexandria, Falls Church, Georgia, Emerson, Temple, and District telephone exchanges in the Washington metro area had all crashed. At no time during the war, even during Pearl Harbor, had whole sections of the Washington phone system broken down. But on that cool night in February 1946, what would later be described as "the most spectacular example of clogged telephone wires this unhappy capital has ever witnessed" took upwards of two hours to clear.

A subsequent reconstruction of this unfortunate state of affairs revealed that within minutes of the announcement, trunk lines leading into the affected exchanges had been tied up by thousands of callers. Not only were umpteen thousands desperately seeking to call the shoe store, but thousands more kept trying—and failing—to reach their aunt Sadie or cousin Ruthie to pass on the word of the offer. Only after the phone company implored the store and radio station to cancel the sale could order even begin to be restored.

Valentine's Day, February 14, 1946, was a day that would live in synthetic infamy. It marked the peak of postwar nylon hysteria, an artificial condition created by the deliberate hoarding of greedy suppliers, who euphemistically admitted to "delaying sales in the interest of 'orderly' marketing." Though *Business Week* predicted as of August '45 the imminent advent of "Yuletide nylons," Christmas came and went, but the women of America went without.

In New York, a week before Valentine's Day, an estimated *thirty thousand people*—"including many brave men," according to *Newsweek*—mobbed the fifth floor of Gimbel's in fierce competition for an advertised twenty-six thousand pairs. In Omaha, five pairs clinched the purchase of a carload of scarce wheat. In Chicago, a furious cab customer sued the Checker Cab Company for $75 ($25 plus $50 "for expenses") when sparks from a driver's cigarette burned a hole in his wife's . . .

The week after Valentine's Day, a reporter for *The New Yorker*'s "Talk of the Town" section bravely joined one of the local nylon lines, finding it "giggly, though with a strong male representation." The door to the besieged establishment had been locked against possible gate-crashers and manned by a formidable-looking uniformed security guard. Inside the hosiery department, a "languid, well-stockinged lady" explained the lengths to which unscrupulous women were going to get their legs on more than one pair. One woman turned up wearing a large floppy hat, demurely carrying a hatbox, bought a pair of nylons, darted around the corner, changed hats, and brazenly rejoined the line. "She went through three hats before we nabbed her," the saleswoman muttered indignantly. "Who knows what they'll think of next?"

What they thought of next, in less decorous sections of the country, was brute force and intimidation. In one midsized Georgia city, the headlines blared: "Women Risk Life and Limb in Bitter Battle over Nylons," after an orderly nylon line degenerated into a distinctly unladylike riot. Like football violence in later decades, nylon riots prompted a flurry of hand-wringing from pundits.

From the sound and fury of the nylon lines, a war-battered silk industry could read the writing on the wall, foretelling its imminent decline. "Spurned Worm," *Business Week* gloated, reveling in the fact that "the mechanical spinnerets of Du Pont have vanquished

Japan's once-haughty silk worms." But in a twist of fate typical of the chaotic postwar economy, now that Gen. Douglas MacArthur's American occupation government was in charge of rebuilding Japan, the American government had been counting on rising silk sales to help pay for its old arch enemy's from-the-ground-up reconstruction.

In 1947, the War Department joined forces with the International Silk Guild in backing a multimillion-dollar promotional campaign to "convince the American public that silk is the superior textile, with peculiar virtues which justify its higher costs." Though the synthetic fiber industry in general cried foul, Du Pont was too busy to protest.

Du Pont had bigger fish to fry than stockings. Dresses, for example—the fashionable kind. And possibly even men's shirts. Unfortunately, the problem with nylon as a luxury dress or shirt fabric was that it was either too stiff and hot for hot days or clammy for cold. Though Du Pont took a stab at selling a shirt-quality fabric to menswear manufacturers, the failed effort badly undermined nylon's deluxe reputation. "Many a disappointed purchaser of nylon shirts," ran one conciliatory advertisement from prominent manufacturer Cluett, Peabody & Co., "maintains that a lack of necessary ventilation made their shirts hot in summer and clammy in winter."

After some shady lingerie makers fashioned a load of military-surplus parachute fabric into women's underwear—ouch!—one industry spokesman decried that "the magic name nylon was being used to promote absurd blends and even outright misrepresentations." Despite such avowals, *Consumer Tests* dashed nylon's last hopes for conquering men's "shirtings" by revealing that a standard machine turbowash resulted in a tattered mass of "split seams, frayed button-holes, yellowing of fabric, and a general appearance which didn't compare well with a standard cotton shirt."

Wash 'n' Wear

February 1948: Dr. William Hale Charch (the brilliant chemist famed for developing moistureproof cellophane) sent two of his top scientists with two 200-pound samples of Fiber V (Du Pont's new top-secret polyester fiber) up to the worsted division of Pacific Mills in Lawrence, Massachusetts, to be processed into staple.

In subsequent tests, polyester Fiber V was judged "extremely resilient" with "excellent resistance to and good recovery from wrinkling." It even displayed "a liveliness and springiness similar to wool." This was music to Charch's ears, because ever since his arch rival, Wallace Hume Carothers, gained immortality for displacing silk, Charch had been on a rampage to replace wool.

Polyester fiber looked like the answer. But just as victory was in his grasp, those idiots at Pacific Mills—without checking with him first—had gone and taken out all the crimp. Trying to make synthetic wool without crimp, in Charch's opinion, was a little like trying to make synthetic silk without sheen.

Charch was further dismayed that the chemists working for him had committed the cardinal sin of mechanically imposing the crimp on the fiber *after* it was cold-spun instead of building it into the molecular structure itself. A better way, a more honest way, Charch believed, was to build crimp into a fiber from scratch. But Charch found himself overruled by the influential head of nylon research, who failed to share Dr. Charch's strong views on the subject.

Never mind—because crimp, or the lack of it, turned out to be the least of Dacron's problems. The first batch displayed disturbing hole-melting tendencies from cigarette burns. It had a bad case of static cling. Even worse, it pilled. In a panic, Du Pont's technical division set up a special lab charged with attacking these problems on all fronts.

They started with pilling. Rival fabrics' tendency to pill—or not pill—was extensively analyzed. The obscure mechanism behind pilling was probed. Finishing techniques that promised pilling reduction in the fiber's final form were tested. At long last, a polyester fiber with a ribbon-shaped cross section—after being road-tested in a pair of slacks worn upwards of fifty times by a human man—appeared to only minimally pill.

With the static cling, pilling, and the cigarette hole meltdown under control, Charch considered Dacron home free. But that was before the ribbon-cross-section fiber revealed a unique drawback: When dyed in dark colors, it tended to glitter garishly in sunlight, like a mobster's moll in the cold light of dawn. It took Du Pont's new Dacron research laboratory three years to solve Dacron's "Las Vegas problem," but once the sheen was swept off its surface, Dacron was ready to roll—or rather, be rolled around in.

At its launch, Dacron was a runaway hit in menswear, because it promised the wash-and-wear moon: drip-dry suits that even after going through the hopper didn't look like something the cat had dragged in. If it couldn't exactly be wiped clean with a damp cloth, Dacron could be machine washed as opposed to dry-cleaned. On the strength of anticipated dry-cleaning savings alone, Dacron's price point zoomed past wool. At New York's Witty Brothers, where a Dacron suit sold for $82.50—close to three times its wool equivalent—sixteen thousand Dacron suits sold in the spring of '52. Proclaiming "A Synthetic Surge," *Time* warned an "already quaking" wool industry: "Look sharp, lest [you] go the way of silk."

The Nicest Thing That Could Happen to Your Kitchen

On the last day of the meeting, a light airplane swooped down out of a clear blue Florida sky and began ominously buzzing the palm-shaded grounds of the Tupperware Corporation. After executing a tight roll over Tupper Lake, the pilot released a confetti-like cluster of cards, which fluttered groundward over the carefully coifed heads of twelve hundred housewives gaily gathered on the greensward below.

On every one of those lily white cards, raining down like manna from heaven, was emblazoned the name of the seventh winner. As the first of the cards landed lightly on the lawn, hundreds of women hiked up their billowing skirts and raced toward the first landfall, each wondering whether she might be the One chosen by fate (and by Mr. Tupper) to be so honored. And the winner was . . . Mrs. Mabel Best of Kansas City!

After the hubbub—spontaneous hugs, showers of kisses from well-wishers—died down, the loyal staff gazed toward the podium in rapture as Earl S. Tupper, founder of the legendary Tupperware empire, personally congratulated Mrs. Best on the occasion of the guaranteed fulfillment of her fondest fantasy.

Which, to put it modestly, represented the lucky chance of a lifetime—a dream nearly forgotten by the time Mr. Tupper plucked it at random from the hundreds of entries submitted in response to an urgent request from Tupperware headquarters.

To her superiors, Mabel Best had earnestly confided her greatest wish—to be reunited with her son Donald, a marine sergeant serving in Japan, on the occasion of his twenty-second birthday. And now, lo and behold, here was Earl S. Tupper himself, publicly announcing that not only would she be treated to a weeklong reunion in Tokyo (including enough birthday cake to feed Donald's entire battalion) but that she and her darling Donald would soon be enjoying, at company expense, a first-class sight-seeing tour of Japan.

Ten years of peace and prosperity had been awfully good to Earl Tupper. And it pleased him to reward the elite of his ten-thousand-strong sales force with a weeklong bonanza of lavish gifts organized around the timely theme that yes, "Dreams *Do* Come True." The night before, Mrs. Doris Stewart—to the accompaniment of oohs and aahs from the crowd—had seen her name spelled out in flashing fireworks across the night sky. This had been Mr. Tupper's clever way of alerting her that she should start pinching herself, because she would be flying off to Disneyland any day now—with family of four in tow.

It certainly seemed fitting that the pink Cadillac bestowed on Peggy Allison, that the spring wardrobe lavished on Anne Carter—who after sitting through an hourlong fashion show had been informed that the clothes on the runway were hers—should be grounded in a postwar plastic prosperity. Because Earl S. Tupper, a self-made, self-educated man, tough as a tumbler, had proven that only in America could a poor boy grow up to be a millionaire after persuading half of America—the better half—to fall in love with polyethylene's "fleshy feel."

Back in 1942, Earl S. Tupper, a thirty-five-year-old self-described "ham inventor and Yankee trader," had proudly declared himself "president, treasurer, and sole stockholder" of the Tupperware Corporation, an enterprise firmly dedicated to turning out "tomorrow's designs with tomorrow's materials." During a brief stint as an engineer at Du Pont in the early forties, he had become familiar with plastic—in particular, with polyethylene—then being manufactured by Du Pont under license from I.C.I. in Britain.

After branching out on his own, Tupper was frustrated by industrial materials being kept strictly off-limits to all but military suppliers. Fortunately, his old bosses at Du Pont were willing to let him have a few tons of polyethylene slag for a song; they had no idea what else to do with it. Growing up on a farm in rural Harvard, Massachusetts, young Tupper had honed his trading skills after learning the hard way, according to *Time*, "that he could make more money buying and selling other people's vegetables than by raising his own." Now a man with a mission, he spent every free moment in his tiny Farnumsville, Massachusetts, factory refining a process that would overcome the paraffin-like substance's tendency to split, while keeping it tough enough to withstand almost anything except knife cuts and near-boiling water.

Late in the war, Tupper Plastics enjoyed a modest success selling polyethylene gas-mask and signal-lamp parts to the navy. But Tupper's big break came in 1945, when polyethylene began trading on the open market at an astoundingly low 44 cents a pound.

To distinguish his private polymer from the run-of-the-mill goo, Tupper called his "Poly-T"—short for "Polyethylene-Tupper." But what mattered more than any putative polymer engineering on Tupper's part was that by late '47 orders for Tupperware were pouring in from clients all over the country: from the American Thermos Bottle Co. for 7,000,000 nesting cups; from Canada Dry Ale for 50,000 bowls to sell with beverages; from Tek Corp. for 50,000 tumblers to sell with toothbrushes; from Camel for 600,000 cigarette cases. Even New York's Museum of Modern Art clasped Tupperware to its stark, unadorned bosom by including two Tupperware bowls in a display of "useful objects" executed with an eye toward good design.

Until Tupperware, American kitchens had been stocked exclusively with glass and ceramic containers, which were easily broken and far from airtight. Tupperware changed the face of leftovers forever by offering the busy housewife an idiotproof, childproof table service, so indestructible that a Massachusetts insane asylum found it an ideal replacement for its battered aluminum bowls and cups, because inmates could destroy their brand-new table service, *Time* solemnly reported, only "by persistent chewing."

Tupperware's vast mass appeal stemmed from a combination of incredible toughness with an almost sensual softness. Rather than picking a hard, heavy, rigid, Bakelite-like plastic in an attempt to simulate the look and feel of ceramic and steel containers, Tupper blatantly reveled in his Poly-T's protean form, its pure Platonic plasticity.

Before the forties were out, Tupperware bowls, tumblers, Econo-

Canisters, and covered butter dishes were winning high marks from highbrow aesthetes for their stark, utilitarian design. In a budget bid to ensure that form followed function, Tupperware butter dishes were shaped like sticks of butter, cake canisters like cakes, pitchers like sleek-spouted carafes. In 1948, *House Beautiful* devoted a multi-page four-color feature to Tupperware entitled "Fine Art for 39 Cents." An ad from the same period hailed the debut of the Tupperware tumbler as "one of the most sensational products in modern plastics," gorgeous in a mouth-watering array of hues, from "frosted pastel shades of lime, crystal, raspberry, lemon, plum, and orange to ruby and amber."

Tupper skillfully exploited plastic's inherent strengths: its intrinsic color and unbreakability. He reinforced the legend of Tupperware's impregnability by offering lifetime guarantees against chipping, cracking, peeling, and breaking. But Tupper's true genius consisted in capitalizing on plastic's friendliest qualities—that Tupperware could be squeezed and fondled into submission; that all Tupper bowls possessed the capacity to "form a spout which disappears when the bowl is set down."

What ultimately set Tupperware apart from the rest of the plastic pack was its exclusive Tupperware seal. All Tupper containers had to be "burped" like a baby, so forming the patented "airtight, liquid-tight" hermetic Tupperware seal.

Tupper had keyed into the collective unconscious by injection-molding a baby-boom product par excellence: supremely safe, posing no threat whatever to kids in the kitchen, and in a pinch, eminently suitable for double duty as a chewable, throwable, bashable toy. On a more subliminal level, at a time when fallout shelters and missile silos were being hermetically sealed off from the tainted air around them, Tupperware vigilantly protected vulnerable leftover food from all external threat. Banishing germs or Germans—or more metaphorically—mad Russkies hurling hydrogen bombs.

Tupper's tell-tale burp had only one major drawback: that satisfying snap, like a paint can in reverse, could not be conveyed to a potential customer without a trained Tupperware team member performing a live, hands-on demonstration. In 1950, frustrated by faltering sales and convinced that conventional retailers were failing to capitalize on the burp, Tupper yanked Tupperware out of retail outlets and turned the marketing program over to Brownie Wise, who had done well by herself personally demonstrating the signature seal at house parties in her home.

As chief operating officer of Tupperware House Parties Inc., exclusive distributor of Tupperware products, Brownie Wise catapulted the plastic container into the cultural stratosphere by institutionalizing a classic postwar social activity: the Tupperware house party. Tupperware parties superficially operated as quasi-social gatherings, the ideal way for a young, socially self-conscious housewife to widen her claustrophobically small circle of friends.

More covertly, they gave thousands of hardheaded fifties housewives with an insatiable zest for home economics an opportunity to safely fuse entrepreneurialism with the traditional social obligations of home entertainment—and home improvement. Tupperware dealers staged decorous socials in the homes of genial "hostesses," who agreed to turn over their chipper, brightly hued Formica kitchens to a dozen of their closest friends in return for free Tupperware.

As the Tupperware empire grew by leaps and bounds—1954 sales topped $25 million—and relocated to sunny Florida, its far-flung network of hard-driving dealers grew accustomed to being showered with lavish gifts, TV-game-show style, at elaborate year-end extravaganzas held at the luxurious Tupper HQ. At its peak, the Tupperware world was a secret society without men (with the notable exception of Tupper), a place where women of skill and ambition could be hardheaded and practical—while, like the products they sold, maintaining a placid, pliant, pious exterior. Like his contemporary Hugh Hefner—who enjoyed playing the consummate host at Playboy house parties—Earl Tupper's world oozed with sexuality, though in Tupper's case it was a repressed kitchen kinkiness, an object-fetishism gratified by lavish gifts of girlish goodies—pink Cadillacs, mink stoles, diamond bracelets—bestowed on the best Tupperware Ladies by Big Daddy. In 1958, after selling out to Rexall Drugs for a reputed $10 million in cash, Tupper became a citizen of balmy Costa Rica—a notorious tax haven.

Better Living Through Chemistry

By the mid-fifties, Americans were consuming plastic in vast quantities, far exceeding the highest peak of wartime. The plastic conquest of kitchen and dining room, den and diner, hotel bar and ocean liner, could in retrospect be considered one of the greatest cases of swords being beaten into plowshares in the history of the boom-bust, war-and-peace capitalist economic cycle.

Having endured combat, plastic was a hardened veteran. But like all those men who'd come home and turned into pipe-smoking genial hosts, weekend barbecue warriors, combative lawn mowers and swimming-pool maintainers, plastic had gone the peacenik route with a vengeance.

Plastic endowed even the most utilitarian objects with zest and spark. Zippers, for example, were made out of nylon, which was impervious to laundering heat and hot water or dry-cleaning fluids and able to withstand temperatures topping 200 degrees Fahrenheit. Waterproof watch straps were polyethylene. Bassinets in transparent Lucite. Convertible hardtops fabricated from "Plexi-Top Full-Vision Plexiglas—bound to win enthusiasts among those who enjoy seeing the countryside at all times regardless of rain, snow or gusty winds!"

Every statement about plastic in the postwar era was concluded by an exclamation point. Because babies now nursed from plastic bottles equipped with collapsible polyethylene film inserts, it was the era of the Burpless Baby! "Joy riding becomes reality in this new Chrysler upholstered in Saran fabric, well adapted to open cars because it resists rain and sun!" crowed a typically exuberant Dow Chemical ad.

The first plastic-shelled TV, by Admiral Corporation of Chicago, boasted a case that weighed *only ten pounds*! An "unbreakable" Plexiglas screen covered the picture tube, providing that all-important plastic protection. The advantages of synthetic materials—ruggedness, ease of maintenance, imperviousness to spills, burns, and hard knocks—remained unquestioned by a public eager to live it up in the living room without having to worry about knife scratches, stubbed cigarettes, cocktail splatter, and the occasional kick in the screen.

Even the word *plastic,* for decades suppressed as a consumer no-no, made a comeback as a "sell" word in the fifties.

"This beautiful palm-of-your-hand Emerson Radio receiver comes in plastic in a wide range of colors."

"The new Eversharp plastic barrel comes in a wide choice of colors, and is virtually indestructible."

"These Barclay Manufacturing plastic-coated panels go up in a jiffy, immediately bringing the richness of warm color into your home!"

Everything had to be done in a jiffy, in the blink of an eye, at the snap of a wrist, at a fingertip's touch—with no muss, fuss, or bother. Plastic helped make it happen—fast. At the office, a title on the door rated a Bigelow on the floor. At home, groceries became easier to load into the back of the station wagon—and so much less fearful to drop!—with the advent of the polyethylene squeeze bottle. In 1952, with twenty-three million of them in circulation after only two years, *Modern Packaging* ascribed the squeeze bottle's meteoric rise to the top of the charts to the fact that it was "unbreakable, inert, tasteless, odorless, and non-toxic." It was also a proven waste beater—squeezed from a nozzle, condiments could be trained at a hot dog or burger with relentless accuracy.

Chores, chores . . . what a bore! Critical as they were to the maintenance of domestic well-being, any plastic product that promised to put an end to a chore, or cut one down to size, enjoyed pride of place on the postwar menu. Though washing walls had hardly been a central activity before the war, in the postwar era of obsession with the elimination of dirt—particularly the "ground-in" kind—the vinyl industry had a hit on its hands with "easy-to-put up, self-applied, adhesive-backed vinyl wall coverings that make keeping walls clean easy as pie!"

For that traditional look, you had your choice of simulated brick, and the ever popular simulated grass fiber. For bathroom makeovers, plastic "ceramic" tile. For lumpy walls, vinyl wall coverings were backed with a thin layer of vinyl foam, so no one—certainly not unsuspecting houseguests—would be the wiser! United Wallpaper introduced a successful line of vinyl-coated wallpapers that precisely matched their vinyl-coated upholstery fabrics: both deeply embossed. In 1956, Carvel Ice Cream, the popular dessert that emulated plastic, made a bold switch to disposable vacuum-formed dishes for sundaes and banana barges. The thin preformed polystyrene sheet proved so successful that waxed paper dishes were consigned to the dustheap of fast-food history. In the first year, Carvel customers consumed a million vacuformed banana barges and 1.5 million sundae dishes.

As *Modern Plastics* politely pointed out, "Though they fall into the category of disposables, their potential reuse value makes them extremely attractive to the dessert buyer for future use in stowing small household items."

Saran Wrap

"How did we *ever* get by without Saran?" gushed *Good Housekeeping* in November 1957, innocently recalling those desperate days when the American housewife possessed no transparent film other than cellophane in which to wrap her precious leftovers. Though Du Pont didn't care to admit it, Dow Saran Wrap had it all over cellophane. Not only did it cling to itself, but it tenaciously clung to just about anything it was wrapped around, including a ceramic bowl or plate. The wonderful thing about Saran was that, like disposable Tupperware, it offered airtight protection. In case you were wondering, that meant "total freedom from germs and alien odors."

Saran had been poised to take the automotive upholstery-covering market by storm until its bid was derailed by unpleasant rumors of excessive abrasion, particularly with regard to women's delicate fur coats. *Modern Plastics* swiftly leapt to Saran's defense: "It is hoped that the absurd fallacy that Saran upholstery breaks the bristles on a lady's fur coat when she moves about on the automobile seat has by now been knocked into a cocked hat. Ladies have been sitting for years on fiber-woven auto seat covers without serious incident. Besides, few women have the heebie-jeebies so severely that they have to keep moving constantly back and forth on the automobile seat."

Whatever the merits of Saran as automotive upholstery, it was in film form that "vinylidene chloride" attained iconic status in postwar America. Long considered a likely candidate as a food wrap, for years Saran's entry into the fray had been hampered by a major defect: It smelled to high heaven. By the mid-fifties, discreet announcements began appearing in the trade press to the effect that "that old bugaboo, unpleasant odor, has at last been eliminated."

This deodorization—the fifties, as a rule, was not a period partial to odors—made Saran safe for food products, and as such it made a glittering debut in the spring of '54 as "the new wrapping material that clings to bowls, pans, dishes and to itself without requiring string or rubber bands to hold it tight."

BIG CHANGE
IN SARAN WRAP!

Announcing... exclusive new package improvement!
The wrap that's best for protection is now the easiest to use!

Courtesy of Dow Chemical

Banishing string and rubber bands from the kitchen was a feat Tupperware had accomplished some years before. But the new "self-clinging" Saran offered much the same ability to banish noxious odors, without *even having to take the food off the plate*! What truly took Saran over the top was that innovative cardboard case, equipped with its own "cutting edge." It just about cut itself!

To give Saran its final push into cultural history, Dow single-sponsored a TV drama, *Medic,* scheduled directly opposite *I Love Lucy,* the most popular show on the air. "Brought to you exclusively by Saran Wrap."

The TV commercials would consistently close with the slogan "the Film of One Hundred and One Uses." Taking that premise literally, *Good Housekeeping* gladly enumerated a few other uses for Saran, other than covering a plate of leftovers and shoving it into the Frigidaire. Saran eliminated not only a vast array of household demons but disposed of worry—sent it right down the drain!

Worried about "freezer-shelf stick," when your ice trays stick to the shelf? "Just cover the shelf with Saran, and ice trays slide out with a finger-tip's touch." Dread dust settling all over your "best company" china? Just wrap each special piece in its own gleaming piece of Saran, and carefully place small squares of Saran as dividers between every plate to add protection from germs. Worried about tarnish on your "best company" silver? Don't worry. Just make sure that "the cling-wrap seals tightly to keep that no-good oxygen-filled air out."

Even worse than no-good oxygen-filled air were messy houseguests who carelessly spilled sauces and dips all over your buffet table. But with Saran, "Just place small squares of Saran under bowls of dips, sauces and other spillable foods and relax." Worried about grease and cooking stains spoiling cookbooks lying open on the stove? "A small square of Saran laid over the open page of a cookbook lets you read your recipe without worry." Saran was its own therapist.

The Surface with a Smile

"Whenever I think of surfaces," a perky young housewife with a Bakelite beehive and frilled apron mused in a pastel-hued ad in *Ideal Home* magazine, circa 1953, "I think of Formica." That gorgeous "wipe-clean" surface, the ad sweetly cooed, was "the nicest thing that could ever happen to your kitchen."

"Won't *you* feel life is good when *you* own a kitchen where all the surfaces are jewel-bright, clean-at-a-wipe Formica Laminated Plastic? Ask for it by name—pronounced 'Formica' as in '*For My Kitchen*.'"

When first manufactured in 1913 by two young Westinghouse engineers who had worked closely with Leo Baekeland perfecting Bakelite laminated, "Formica" had stood "for mica"—the costly mineral used in electrical insulation. After failing to persuade Westinghouse to form a division to manufacture Bakelite electrical products under its own name, Daniel J. O'Connor and Harold A. Faber formed Formica Corp., headquartered in Cincinnati, Ohio.

Alarmed that Formica might one day compete with Bakelite laminated, Baekeland promptly slapped Faber and O'Connor with a patent suit. But having to defend themselves from the famously litigious father of plastics only endowed their tiny start-up with the image of a scrappy contender. After switching suppliers to Baekeland's arch nemesis Redmanol, Formica sales soared on the radio wave of the twenties and thirties. During the depression, Formica patented a method of producing a wide range of decorative effects by printing a top plastic layer on a lithographic press, making it possible to uncannily simulate wood grain and marble veining—the area in which Formica would shine above its rugged rivals.

Like its cousin Bakelite, Formica made its first conquests not in the home but in institutional settings, where a remarkable ability to withstand punishment compensated for any disconcerting lack of authenticity. Hotels, soda fountains, restaurants, roadside diners—establishments where surfaces took extraordinary abuse as a matter of course—became Formica's earliest satisfied customers. The Library of Congress installed wood-grained Formica sheeting on its library tables. Washington's Statler Hotel lined the walls of its lobby and elevators with it. The Cunard liner *Queen Mary* paneled its state-rooms with Formica. After taking five years of hard knocks during its tour of service as a troop ship during World War II, the big boat sustained no more than minor damage.

After earning a tough-as-nails reputation during the war (where it loyally served as paneling for prefabricated military barracks and hangars) Formica came into its own in the fifties as the bright, gleaming new tabula rasa America yearned for to make a fresh start after fighting that filthy war. It was not for nothing that it was called "the wipe-clean wonder," because at the whisk of a damp cloth the unruly past was swept away.

Hard-edged, bold-colored plastic laminates were perky, clean, and *nice*. With so much uncertainty threatening the domestic sphere in an era of atom bombs and covert left-wing subversion, stolid Formica took a toughness tempered on the beaches and in the jungles of combat and turned it to good use banishing the banes of any housewife's daily grind: dirt, juice, and alcohol stains, and the ultimate social faux pas: cigarette-butt burn. In those hard-drinking fifties, Formica counters spent the night shift fending off the wayward attacks of obstreperous houseguests, while doing day duty shrugging off the random bashings of rambunctious kids. Uncanny simulations of birch, brown stump walnut, maple, mahogany, and other rare wood veneers let Formica accomplish a second goal of the suburban dreamer: jumping a rung up the social ladder by endowing a strictly middle-class ranch house lacking "background" with an elegantly appointed "wood"-paneled library or den at a cost not much greater than a fresh coat of paint. Formica could install tradition in the dining room while evoking the Space Age in the kitchen.

Formica's wide range of colors and patterns permitted happy homemakers to indulge in a modicum of rugged individuality without trespassing the confines of the prevailing conformity. As *Fortune* put it, "Chances were your Formica kitchen wouldn't look just like the one across the street." By choosing wisely, one could keep up with the Joneses while simultaneously differentiating oneself from them. When Formica hired Raymond Loewy to create a new line of laminates, he uncannily incorporated the now classic "boomerang" shapes, floating on white and pastel fields, which Loewy himself referred to as "atomic" or "biomorphic." Ironically, the "wipe-clean wonder" that alleviated anxiety by bestowing "the surface with the smile—the nicest thing that ever happened to your kitchen" also provided a blank screen for the unconscious projection of a prime fifties anxiety: the atom bomb. Just as Tupperware offered hermetic protection against a potentially poisonous atmosphere, Formica provided protection against internal and external attack, eternally vigilant in its struggle to wipe clean the past.

From Bombers to Boudoirs

Three months before VE Day, *Life* charitably devoted a multipage spread to Rohm & Haas's desperate attempt to promote Plexiglas as "one of the plastic substances which designers love to put in their glittering postwar plans." In the eyes of its inventors, if not its beholders, Plexiglas was slated to be the postwar world's glamour plastic, the contemporary equivalent of old-world crystal.

The end of the war was a tough time for Plexiglas. Rohm &

Haas founder and patriarch Otto Haas was forced to testify before Congress to defend himself from an antitrust suit filed by the Justice Department, accusing him of conspiring with Du Pont (maker of Lucite) to fix the price of acrylic plastic in violation of the Sherman Antitrust Act.

By August 14, 1945—VJ Day—Plexiglas production had plummeted to an all-time low. Rohm & Haas's magnificent "three-room Plexiglas Dream Suite" represented a last-ditch effort to reshape public opinion into a pro-plastic frame, just as its authors had once blow-molded bomber blisters.

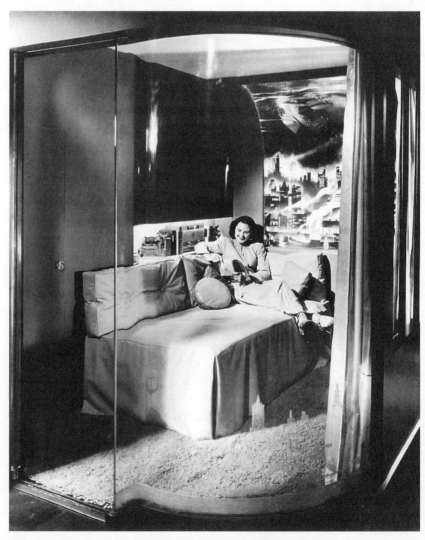

Courtesy of Rohm & Haas Company

During its three-week installation at Wanamaker's Department Store in downtown Philadelphia, the Plexiglas Dream Suite attracted some fifty thousand visitors, all visibly wowed by the sexy glamour of a "transparent" bedroom, which directly adjoined a bathroom equipped with a turret-shaped transparent stall—ooh, aah! In the bedroom, a Plexiglas "lazy Susan" revolving hat rack hung above a row of Plexiglas shoe racks. A four-color mural over the bed was "painted in light" by the phenomenon known as "edge-lighting," a property unique to acrylics by which light could be "piped" around corners, permitting the short-lived design innovation Rohm & Haas dubbed "radiant walls." Even the handles on the faucets in the transparent bathroom were Plexiglas—"indelibly marked 'hot' and 'cold' with dots of color." The toilet, of course, topped by a transparent seat, was discreetly opaque—no designer's glittering postwar plans included a clear-eyed view of that most dreaded of organic substances. Plexiglas, after all, was meant to evoke the ethereal, not the corporeal.

While veterans with a nostalgic attachment to Plexiglas-blistered airplanes eagerly snapped up acrylic cigarette boxes and cases, umbrella handles, and costume jewelry for their wives, the once mighty shatterproof crystal ball of the skies didn't take long to be relegated to the status of cheap novelty. Though the Plexiglas Dream Suite toured twenty major department stores nationwide, by the late forties the short-lived consumer boom in acrylics had gone bust. The one purpose left for poor Plexiglas was to be molded into red-and-amber-colored lenses for auto taillights.

Penta-Plastic

February 1949: On the lawn surrounding the Pentagon, a bizarre little package awaited the casual inspection of the nation's high armed forces officials. No one was quite sure what to make of it, or even quite what it was. Looked at one way, it appeared to be nothing more than a pile of tubular steel beads with an internal steel cable randomly wending its way through a spaghetti-like latticework. But, as its eccentric inventor was willing to demonstrate upon request, all you had to do was tug on that cable and—presto!—from a suitcase-sized package an entire round room sprung into view. Or, looked at from a different perspective, a very compact plastic house.

As R. Buckminster Fuller—the Harvard dropout who dreamed up the outer-space creature called a "geodesic dome"—would patiently explain to willing listeners, this unique structure utilized an arcane geometry all of its own. Triangles and tetrahedrons formed trusses capable of distributing weight radially outward, enabling the "outwardly tensed" skin of the building and its supporting structure to be lighter—yet stronger—than any other building type.

As a way of promoting his structures' strong suit, Bucky was often moved to ask, rhetorically, "Madame, do you know how much your house *weighs*?" Madame most assuredly did not. But it became Fuller's mission to persuade people that the question mattered, and that with its solution untold benefits would be reaped by those who knew how to radically reorient structures from the top down.

Having grown up around boats on his family's private island off Maine, Fuller's idiosyncratic style of architecture seemed like an effort to make buildings that floated boatlike on the earth's surface as opposed to anchored in the soil like rocks. From his early days sailing catboats in Bar Harbor, Fuller had been obsessed by speed. In the late twenties, in his twenties, inspired by the "streamlined" designs of the Art Deco era, Bucky devised a streamlined, snub-nosed, lightweight, aluminum-clad, three-wheeled vehicle capable of turning on a dime—Fuller described it as "omni-directional."

Even the car's name—"Dymaxion"—was a classic Fuller invention, standing for "Derived Maximum Output from Minimum Materials." Following this concept, even a house, by Fuller's lights, should be free to move like the wind. For years, he dreamed of a house that could be airlifted into place by helicopter. A house that, like a glider, would effectively be lighter than air.

After proudly pronouncing the Dymaxion car a conceptual success despite its commercial failure, Fuller turned his sights on houses. After the tragic death of his daughter Allegra—from a variety of causes, including pneumonia, which be believed had been brought about by a lack of ventilation—Fuller developed his Dymaxion Deployment Unit, a self-contained, self-cooling cylinder with a corrugated aluminum skin that made it resemble an oversized oil drum.

During the war, Fuller managed to sell a few Deployment Units to the army, which housed defense workers in them. But the Dymaxion Deployment Unit was merely a dress rehearsal for the geodesic dome, which fulfilled the promise laid out by the utopian charter of the Fuller Research Institute (established 1946): "To conceive and implement Commonwealth pertinent, individually conceived, intuitively urged and spontaneously joined search, research and enterprise in the borderline realm of 'just not impossibles.'"

Say what? No matter. Like the rest of America, Fuller had become positively obsessed by plastic. But while middle-class suburban America contented itself with slapping down a Formica panel or counters, like a snake shedding its skin Fuller was intrigued by the idea of radically reducing the frame—and "outwardly tensing" the skin.

For outward tensing, no fabric came better prepared than the super-strong polyester film just coming onto the market. In 1951, Fuller put his "just not impossible" dream to the test by erecting a hyperbolic-parabolic dome skinned with a hypercoat—a tough polyester film—on an island in Labrador's Baffin Bay. After withstanding six months of arctic winter conditions without any detectable decay, architectural experts and military strategists who had formerly dismissed Fuller's domes as a cheap fad were forced to sit up and take notice.

January 1953 launched the transformation of the "just not impossibles" into the realm of the probable—and profitable. On the eve of the fiftieth anniversary of the Ford Motor Company, Henry Ford's grandson, Henry Ford II, had an inspiration: Why not put a great dome over the Ford Rotunda at River Rouge—a circular arcade of columns constructed of steel beams covered with stucco surrounding a courtyard open to the wind and sky?

The dome would be the ideal monument to his grandfather, who had always viewed the magnificent, open-air Rotunda as a shining symbol of Ford's future. But Ford's in-house engineers, in consultation with some of the world's leading architects, promptly pronounced the project impossible. The most efficient conventional dome capable of spanning the ninety-three-foot-diameter circular court would, they maintained, have to weigh a minimum of 160 tons. Under such a staggering load, the Rotunda's walls would come tumbling down like Jericho's.

At a loss for solutions, one of the consultants suggested they call Fuller. When asked if he could put a ninety-three-foot geodesic dome over the Ford Rotunda, Fuller replied simply, "Yes." When asked how much such a dome would weigh, he shot back (after a few days' delay to crunch numbers) with the precise figure of eight and a half tons—a twentieth of the weight of a conventional dome.

In the face of resistance from a skeptical board, Henry Ford II rammed an unconventional contract through the Ford bureaucracy, granting Ford the right to manufacture a geodesic dome under Fuller's supervision and committing Bucky to having his dome upstanding by the next stockholders' meeting in early April—just four months away.

Fuller built the Rotunda dome from the top down, using workmen positioned on a wide bridge temporarily spanning the courtyard. At its completion, with just two days to go before the stockholders' meeting, Fuller's Ford Rotunda dome weighed precisely what he had promised it would: eight and a half tons, or roughly two and half pounds per square foot of area covered.

When Ford's directors and stockholders arrived to inspect the soaring structure, they were transported by its lofty transparency. Fuller's lacy filigree of trusses, though capable of sending a structural engineer's spirits into orbit, was no more than a frame to the untrained eye. The transparent fiberglass sheeting, however, through which shining sky and floating clouds could clearly be seen, provided a quasi-spiritual experience. No one had ever seen anything like it before—the plastic transparency turned the Ford Rotunda into one of the most popular tourist attractions in the Detroit area.

The Ford imprimatur turned Buckminster Fuller from a poor eccentric into a rich one. In 1954, at Orphan's Hill in North Carolina, Fuller achieved his childhood dream of watching a thirty-foot-diameter plastic-skinned dome be picked off the ground with a helicopter and swooshed through the air with the greatest of ease at a wind-whistling fifty knots. The dome's natural streamlining made it the one structural form on the planet capable of being so ethereally transported. A Marine Corps helicopter airlifted a three-helicopter hangar from the deck of an aircraft carrier to a beachhead three miles away—in a strong headwind—while a company of untrained marines set it up in under four hours.

This little stunt was not just a lark—more than anything else Fuller had done to date, it forced the Pentagon to pay attention. Geodesic domes' unique capacity to be flown through the air and put up in a flash by untrained personnel was of definite interest to the Defense Department. In 1953, the Pentagon began laying plans to build an immense DEW line—the Defense Early Warning radar system—to span the Arctic Circle between Russia and Canada. A direct descendant of Robert Wattson-Watt's 1936 coastal chain around Britain, the DEW line radar stations would need to be protected from fierce Arctic storms by a lightweight, low-cost, extremely strong shelter that would have to be pervious to microwave beams.

Fuller erected a thirty-foot-diameter test "Radome"—constructed of diamond-shaped polyester-fiberglass struts covered with a fiberglass-reinforced polyester skin—in twelve hours, using untrained personnel, and left it on top of Mount Washington in northern New Hampshire for a year. After it withstood howling winds of up to 180 miles per hour without icing up and transmitted radar signals without any detectable distortion, the Pentagon commissioned a few thousand more, in sizes up to three-times that of the test model. That single order made Bucky Fuller a millionaire overnight. Of the thousands of fifty-foot Radomes ultimately strung for five thousand miles along the Arctic Circle, none ever collapsed in a storm. None ever had to be replaced. These were polyester pup tents with the structural specs of the Pyramids.

Plastic—in the form of polyester film—had conquered the elements. For a display designed to promote the durability of corrugated cardboard, the Container Corporation of America—the nation's largest cardboard box maker—sent two large paperboard domes to the Tenth Triennial Design Exhibit in Milan. After attracting tens of thousands of awestruck spectators, a mild northern Italian drizzle threatened to melt them into soggy puddles of paper. Before the twin domes dissolved entirely, an alarmed Container Corporation official placed an urgent call to Fuller at his Bear Island home. Thinking fast on his feet, Fuller proposed that the cardboard shells be covered—as soon as possible—with vinyl "bathing caps." It was a face-saver for the Americans, but not exactly the advertisement for cardboard American Container had bargained for. The cardboard domes looked slightly silly in their vinyl bathing caps—but at least they remained standing. The need for plastic reinforcement underscored the evolving adversarial relationship of plastic to nature: Plastic 1, Nature 0.

The Shell Game

In southern California, people liked plastic for no other reason than that it made an ideal material for fun in the sun—on the road or at the beach. Plastic could be easily turned into surfboards, boat hulls, cute little kit cars, and all sorts of stuff worthy of being rhapsodized about by the Beach Boys.

During the unusually balmy winter of 1950, boatbuilder Bill Tritt was tinkering around his Green Dolphin Boat Works in Costa Mesa, feeling kind of burned out by boats. Things began looking up when Maj. Ken Brooks of the air force station at Long Beach, an old friend, walked into the shop with a personal problem. A few weeks before, Brooks had given his wife a gift that she hadn't much cared for: an olive-drab army-surplus jeep that he had bought for a song and left in the garage, all tied up in a bow. For some reason, he had taken it into his head that a cheap jeep would be the perfect small runabout for his wife, for casual shopping and tooling around. But Mrs. Brooks found the jeep ugly, boxy, and a little too macho for her refined tastes.

"What I need," Brooks sighed, "is a custom-built body. And about $2,000 to pay for it."

"Maybe I could mold you one cheaper," Tritt volunteered, surprising himself by sounding so eager to branch out into the car business, an area in which he had no experience whatsoever. "I'd say maybe $700. If I use plastic."

"Plastic" was the hot-rod word in automotive styling that year. In postwar translation, it meant: cheap thrills. At the time, enthusi-

Copyright 1978 GM Corporation

asm was running highest around Costa Mesa for something called GRP, which stood for "glass-reinforced plastic." It would soon be better known as fiberglass.

GRP was another one of those war babies being promoted as the next miracle material. All you had to do was mix it, mold it, let it harden into resinous form, and presto! You had a swimming pool, a new car, a cabin cruiser. Costa Mesa itself had become a veritable beehive of fiberglass experimentation. The world's first fiber-plastic kit bodies were developed right down the street from the Dolphin Boat Works to fit a Crosley chassis, the brainchild of self-made Cincinnati millionaire Powell Crosley, who had turned a wartime tire liner, radio, and refrigerator fortune into seed money for a small, low-cost sports car. By building kits, young sports-car fanatics could bypass the bureaucracy of the Big Three and reach other like-minded enthusiasts by advertising in the back pages of the hobby magazines. Good-bye, Detroit; hello, Hollywood.

All fired up about building a car after all those years building boats, Tritt cheerfully spent the rest of the winter, and much of the following spring, painstakingly molding a one-piece plastic body to be mounted on Brooks's jeep chassis. Working in his off hours in the dilapidated shed he called a factory, it took him eight months to make a working plaster mold of the silhouette and to cover it with layer upon layer of glass fiber and polyester resin. This was hardly the blink of an eye, but when Tritt was through, he liked what he had done. In particular, he liked that using plastic had let him do it all by himself, with the gratification of fifties ease and convenience. Just as it was going to be easy to clean that all-plastic dream house in a blink of an eye, so any average Joe could now build a car in his garage, even if it might end up taking a little longer than the directions on the box indicated.

Before showing the car to Brooks and his wife, Tritt insisted on taking the jeep-mounted kit car on a ten-thousand-mile test run. Just as Henry Ford had predicted ten years before, the plastic shell was dentproof and squeakproof. Its light weight (185 pounds) saved a fortune on gas, oil, and tire wear. But the most impressive thing about it was its strength: Slammed with a hammer, Henry Ford style, it neither dented, chipped, nor cracked. In a side-on twenty-five-mile-per-hour collision with a pine tree, the only damage sustained was a fourteen-inch hairline crack in the fender.

Mrs. Brooks didn't give a hoot about impact resistance, gas mileage, or tire wear. But she very much appreciated the fact that her brand-new sporty two-seat roadster—which Tritt christened the Brooks Boxer in her honor—gleamed in a pale, leafy green that seemed baked, not painted on. She didn't know, and couldn't have cared less, that an ugly old jeep sat underneath that shiny new shell. It was the sexy, sleek shell that mattered to her: it had a So-Cal surface.

Fortunately for Tritt, Mrs. Ken Brooks wasn't the only California girl or boy to fall madly in love with the Brooks Boxer. After its official launch in late 1952, the Brooks Boxer Supersport, vaguely reminiscent of a Jaguar XK-120, became a runaway hit with California sports-car enthusiasts, many of whose wives, Major Brooks later claimed, "liked to tee off against a garage door."

Tritt was doing a thriving mail-order business when the Connecticut-based Naugatuck Chemical Division of U.S. Rubber—which supplied Tritt with all his resins—offered to buy a batch of Brooks Boxers on one condition: that he supply them with a modified version to distribute under their own name. The makers of the infamous Naugahyde split vinyl upholstery material were looking to get into consumer plastics in the worst way. They renamed the Brooks Boxer the Alembic One—in honor of an old-fashioned beaker commonly used in chemical distilling.

Naugatuck Chemical packed one of their first finished models off to GM in the hope that the sight of the first synthetic car would give the powers-that-be there the idea that a mass-produced plastic car might loom large in America's future. If a plastic car could be taken out of the hobbyist's backyard and onto the assembly line, Naugatuck Chemical would sell a lot of fiberglass.

In the heart of the Motor City, the towering Harley Earl, the automotive genius famed for shaping and styling some fifty million GM cars between 1926 and 1960, strolled out of his office at GM Styling and practically stubbed his toe on the fiberglass Alembic One gleaming on its plastic pedestal in his hallway.

If Harley Earl had done nothing else—and he had done just about everything in the car business—his place in American automotive history would have been secured by his groundbreaking design of the 1952 Buick LeSabre. Earl had boldly pioneered the then radical concepts of tail fins, wraparound windshields, and "Dagmar" bumpers, which resembled modified bomber fuselages.

The '52 Buick LeSabre was a regular B-29 for the road, jam-packed with mechanical references to the great warplanes of yesteryear, aimed at building in a gut appeal to all those ex-fighter-jocks who missed sitting behind wraparound Plexiglas windshields with fingertips poised over a Norden bombsight. Not to mention all the other would-be tough guys who had missed their chance during the war.

Harley Earl was transfixed by the Alembic One. As it so happened, his son Jerry had been pestering him for a cheap roadster to drive to college, and after giving Tritt's plastic roadster the once-over, Earl concluded that he might just have found the answer to his and his son's prayers. Earl had been on the lookout for something new and different to juice up the 1953 Motorama—the traveling road show that previewed new GM cars in major motor markets around the country. This, Earl had a gut feeling, was it: the low-cost, lightweight, stylish sports car young American men were waiting for, without knowing what they were missing.

Too proud a man to knock off some other guy's design, Earl vowed to produce a GM-styled lightweight plastic roadster that would be *the* hit of the upcoming Motorama—possibly of the decade. As chief of GM's legendary styling department, the man who fit the fins on the Buick had long since graduated out of nuts-and-bolts design work. He gladly handed over the pen-and-pad work to Bob McClean, a recent Cal Tech grad and sports-car enthusiast, with strict instructions to conceive and execute a plaster model of a stylish fiberglass roadster to retail at the base price of a Chevy sedan: under $2,000.

According to legend, McClean began by sketching the rear tire, shoving the seats in as close to the rear tire as possible, and then jiggling the standard Chevy stovepipe flame-six engine in as close to the fire wall as practical without throwing the frame out of whack. The end result was a wheelbase of 102 inches. And a price tag to match. The bottom line was that Bob McClean couldn't pass up the chance to design a true American sports car in the grand European tradition.

In May 1952, Harley Earl presented a full-sized plaster model of the Motorama Sports Car—as it was then called—to Ed Cole, the flamboyant new chief of GM Engineering. Much to Earl's relief, Cole was not in the slightest bit put out that the car would have to retail in the neighborhood of $3,500, not $1,800. Not only that, but after literally jumping up and down at the unveiling—a rare case of true love at first sight—Cole urged Earl and McClean to spend the next few months brainstorming for a name.

Only at the last minute, just days before the launch, "Corvair" was rejected in favor of "Corvette." The reason? Earl firmly believed that selling cars to the younger generation would always benefit from a healthy dose of wartime nostalgia. The Corvette, as a GM press release patriotically put it, recalled "the trim, fleet naval vessel which performed such heroic escort and patrol duties during WWII."

The prototype Corvette that debuted at the '53 Motorama came equipped with a lightweight plastic body. But hidebound GM Corporate, a bastion of Iron Age consciousness, had insisted that the production model be executed in conventional sheet steel. Only when surveys revealed that the plastic shell had become a major draw for the majority of young, male spectators who hung goggle-eyed around the lobby of the Waldorf-Astoria on New York's Park Avenue, drooling over the little roadster finished in creamy Polo White, was plastic grudgingly reinstated. As it turned out, plastic also saved money: According to Ed Cole, tooling up in sheet steel would have run $4.5 million. GRP cost $400,000.

The Corvette's fiberglass body ended up being supplied not by Bill Tritt, or even Naugatuck Chemical, but by Molded Fiber Glass of Ashtabula, Ohio, which outbid Naugatuck on the first $4 million order. As a sales gimmick, the first Corvette production run was reserved for "VIP Customers Only"— celebrities, socialites, leading businessmen, community leaders, honchos. More than one plumber with the $3,500 price of a Corvette in his pocket was forced to ask his boss to buy his Corvette for him.

After making a huge splash with the public, the Corvette in its first rollout turned out a bomb. What killed the Corvette in its first incarnation was not its plastic body—which buyers loved—but the deadly-dull Powerglide automatic transmission Earl and Cole, in their mad rush to production, had insisted upon installing as standard equipment. Henry Ford's dream of producing a lightweight plastic car had been at last fulfilled. Unfortunately, Ford under his son Edsel was more committed to retrograde steel. In 1954, Ford struck back with its sporty all-steel Thunderbird. It outsold the plastic 'Vette by a margin of three to one.

Origin of a Species

With plastic skin all the rage, it was hardly surprising that the last refuge of domesticated masculinity, the den recliner, was long over-due for a makeover. Rather than lose the imperial look and clubby feel of top-grain cowhide, Naugatuck Chemical had a vinyl uphol-stery that exuded the raw masculinity obtainable only through trial by fire. If you thought leather was macho, Naugahyde, as Nau-gatuck Chemical crowed, "after being tested under the most severe weather conditions for military use, was adopted in 1943 as mandatory equipment on all U.S. Navy combat ships."

Before the war was out, rugged vinyl upholstery was being used in all types of motorized war equipment, from tanks to trucks to jeeps, and for "wall lining-and-seat covering in bombers, fighters and transport planes." But with all those bombers, fight-ers, and transport planes moldering into rusty scrap, the new mar-ket to conquer was the suburban split-level. In the dimly lit recesses of the wall-to-wall-carpeted den, beneath the sports tro-phies, beside the wet bar, the elusive Nauga could be found in its natural habitat.

> An extensive new line of plastic upholstery for civilian use, to be known as *Naugahyde*, will be made in a wide range of light and bright clear colors and two-tone effects as well as a variety of grains. Waterproof and flame-proof, the new mate-rial, it is claimed, will not get hard or crack, will resist edgewear, abrasion, scuffing, flexing and wrinkling. It will not be affected by perspiration, salt water, alcohol, gasoline, oils, greases, most acids and alkalis; it can be cleaned with soap and water.
>
> —*Architectural Record,* August 1945

The Naugahyde process kicked off in a tiny Mississippi River town near Baton Rouge—Geismar, Louisiana. A joint venture of U.S. Rubber and Borden's called Monochem, the plant turned out the vinyl chloride monomer that would be turned into hairy-chested Naugahyde. The refining process began with the breakdown of nat-ural gas piped directly in from nearby gas fields. After breaking the gas up into its component parts—acetylene, carbon monoxide, and hydrogen—by heating it up to 2,700 degrees Fahrenheit, the acety-lene gas was siphoned off and piped into vats for condensation.

"Making Naugahyde is a bit like baking a pie crust," crowed U.S. Rubber, gamely playing up the down-to-earth metaphor in which the PVC particles were the "flour" while plasticizers were the "water." Mix them together in proper proportions and you came up with a sludgy vinyl "dough" that could be mixed to yield Nau-gahyde soft enough for a baby's bottom, or tough enough for a BarcaLounger.

The dough was rolled between heated rollers, pressed into thin sheets, and coated with a fabric backing. A second set of rollers embossed the molten surface with pebble grain, smooth grain, or no grain at all. If you wanted Naugahyde soft as chamois, you made a sandwich by pressing a layer of foam between the vinyl surface and the cloth backing. If you were looking for a suedelike Naugahyde, you glued fuzzy fibers on it.

> WHEN IS A CHAIR A BEAR FOR WEAR?
> Genuine Naugahyde by U.S. Rubber. This wonderful lightweight comfortable lounge chair in winged Queen Anne style owes its long-wearing beauty to Royal Naugahyde, the vinyl-coated fabric developed by U.S. Rubber not to crack, fade, stiffen, and to take all the wear a busy family can bring to bear. Stays bright and clean with "just a swish of a soapy sponge."
> —1954 AD FOR ROYAL NAUGAHYDE

Vinylmania

The Indian shellac industry had long since lost the electrical appli-ance market to plastic. But now, since the war, it was slated to lose the record industry as well. A *Variety* headline, "War Boffs Disk Biz," said it all. Though a series of scrap-shellac drives—in which music lovers turned in old 78s so they could be recast into new ones—kept the beat going on for the duration, come peacetime the American music industry embraced vinyl as its synthetic savior.

In 1930, the Carbide & Carbon Chemicals Corporation, a large vinyl manufacturer, had asked a team of acoustical technicians at Philadelphia's Mellon Institute to place a few grams of hot PVC (polyvinyl chloride) resin between two record matrices and heat-mold it on a flat platen press.

This test plastic record was played at a secret facility in Camden, New Jersey, for a select group of RCA executives. "The vinyl record," their report conservatively concluded, "seems to have a somewhat lower noise level than shellac, and seems to be quite tough and difficult to break," This was something of an understatement, since compared to brittle, fragile shellac, vinyl was well-nigh indestructible. Though an early generation of Vinylite (Carbide & Carbon Chemicals trade name) records were turned out in the thirties and forties, they were used strictly for electrical transcription purposes and never saw the light of day outside a recording studio.

With vinyl supplies sharply restricted to defense purposes during the war, RCA was forced to hold off until October 1946 to introduce its transparent bright red Vinylite Red Seal twelve-inch 45-rpm recording of "Till Eulenspiegel," at the premium price of $3.50. Sample pressings were circulated to Victor dealers worldwide to drum up support. Hi-fi fanatics, relatively rare in those days, were favorably impressed by the superior tonal quality of the new plastic records. But the general public, even after the price was slashed to $2.50, was underwhelmed by plastic records. With the one notable exception of parents of very young children, who were willing to pay a premium for plastic, not for its obvious sonic superiority but for its safety and durability in the nursery.

While RCA forged ahead with its costly, transparent red vinyl 45s, rival CBS seemed stuck back in the shellac age. But CBS was actually biding its time, waiting for the right moment to strike.

In the fall of 1948, while attending a cocktail party in New York, the Hungarian-born Peter Goldmark—the chief audio engineer at CBS Records and an amateur cellist of some distinction—sat rapturously on a sofa in the living room of a friend's Manhattan penthouse, listening to a superb rendition of one of his favorite works by Brahms: the Second Piano Concerto, in which the third movement has a haunting cello solo. Halfway though the piece, without a moment's warning, his romantic reverie was rudely shattered by a sudden silence, followed by a hideous hissing.

Goldmark knew only too well the source of that vile sonic assault: the shrieking of pickup on paper label, rudely announcing that the side was over. The barbarity of this musical *coitus interrup-*

tus drove the sensitive Goldmark to distraction. How could the recording industry continue to permit this desecration of great art to go on indefinitely? With a sigh of disgust, the well-mannered Goldmark stood up and stalked out of the room, failing to say good-bye to his host.

When calling the next day to assure his old friend that the party had been marvelous, and the musical selection impeccable, Goldmark laid the blame for his abrupt departure squarely on the face of his own industry, which permitted only twelve minutes of music to a side. Such a squashed presentation might be acceptable to the likes of Bobby Darin, but when applied to serious symphonic music, it was an insult to music lovers and audiophiles alike.

Six months later, Goldmark's long-awaited long-playing record, also issued in thick virgin vinyl, came out etched with 250 grooves per inch, a far denser, richer medium than the old shellac standard of 80 grooves. CBS's first LP releases were all personal Goldmark favorites: Mendelssohn's Violin Concerto, Tchaikovsky's Fourth Symphony, and for the middlebrows, Richard Rodgers's *South Pacific.*

For his part, Goldmark couldn't have been happier when RCA (which had turned him down for a job when he first arrived in New York from Hungary) stubbornly stuck to the 45, despite widespread enthusiasm for the long-playing record. As far as he was concerned, it served RCA right to lose the "war of the speeds."

By the time the LP prevailed, vinyl records were faced with a new challenger: audiotape, developed by Marvin Camras, a native of Glencoe, Illinois, and a boy genius had who built a flashlight at four and a functioning transmitter at seven. When he reached his late teens, one of his cousins, who cherished dreams of becoming an opera singer, asked his gifted, mechanically inclined relative to build him a wire recorder so he could hear himself sing arias at home.

Loosely adapting the magnetic-wire technology originally developed by the Danish engineer Valdemar Poulson—who demonstrated his *télégraphone* at the 1900 World's Fair in Paris—Marvin Camras cordially obliged. But in the process of building the recorder, he took into account the later, groundbreaking work of German engineer Fritz Pfleumer, who in 1928 had developed a prototype recording tape made of strips of paper coated with iron oxide.

Camras's contribution to the evolution of magnetic recording tape was to embed tiny metal particles in layers of polyester film, while equipping his revolutionary recording heads with tiny magnets, which rearranged those particles in patterns reflecting the incoming and outgoing waves of sound. The result was a tape recorder that so impressed Camras's professors at the Illinois Institute of Technology that they arranged for him to be hired by the Armour Research Foundation (now the ITT Research Institute), where he could develop it to his heart's content without regard to commercial pressure.

In 1944, Camras's first tape recorders were used for military training, preparing the troops for the upcoming invasion of France. In battle, tape recorders were used to simulate battle sounds at sites where invasion landings were *not* scheduled to take place—creating sonic diversions for the defenders. Camras's tape (which laid the foundation for videotape, floppy, and compact disks) was warmly embraced by audiophiles. But because it was such a pain to handle, it never posed any real threat to vinyl records until the advent of the cassette—a storage system developed by the Phillips company in 1963. Camras's cousin, though he enjoyed his tape recorder, never did make it as an opera singer.

Plast-O-Matic

In the spring of 1930, after completing his sophomore year at Washington University in St. Louis, Charles Eames dropped out of college and hung out a shingle declaring himself a practicing architect—in depression-era St. Louis. When business failed to turn up, Eames took off for Mexico, returning after five years only to accept an offer extended by Eliel Saarinen, dean of Michigan's progressive Cranbrook Academy of Art, to study art and architecture there—on his own terms, free of charge.

After joining Cranbrook's faculty in 1940, Eames joined his dean's son Eero Saarinen—also an aspiring architect and the future designer of CBS's towering basalt black-rock headquarters in New York—in an unusually fruitful collaboration with a young painter, sculptor, and fellow Cranbrook graduate, Ray Kaiser. The three jointly submitted a number of highly original and innovative molded laminated wood pieces to an Organic Furniture Competition organized by the Museum of Modern Art in New York, a competition they won hands down.

With that triumph in hand, Charles Eames and Ray Kaiser married and moved to Los Angeles. By day, Charles worked as a set designer at MGM. By night, he stayed up with Ray until all hours of the morning tirelessly experimenting with new plywood and wood-molding techniques. The Eameses' innovative plywood designs soon caught the attention of the navy, which had fallen in love with fast-bonding urea-formaldehyde resin (a direct descendant of Bakelite) to heat-mold plywood into sculptural forms, molding everything from PT-boat hulls to vast, unsupported airplane hangars.

The new "plastic wood" eloquently spoke to a new generation of marine, aviation, industrial, and residential architects, along with civil engineers and artists, all excited by the prospect of exploring new ways to mold boats, gun turrets, airplane fuselages, and glider shells—on Uncle Sam's tab.

In blacked-out Los Angeles, Charles and Ray Eames played their part by molding splints and stretchers for wounded servicemen. By the war's end, the Eameses' work had attracted so much attention in civilian circles that MOMA dedicated its first "one-man–one-woman" show to a selection of their latest examples of molded-plywood furniture. The public, eager for something different, fell in love with these satiny-smooth forms, which combined the sleek look of Danish Modern with a distinctly rugged American profile.

George Nelson, design director of the Herman Miller Furniture Company, in keeping with the general adoration, commissioned a mass-produced line of molded-plywood furniture to be made exclusively for Miller by the Eameses. As they struggled to adopt their piecemeal process to the mass market, working out of a garage in gritty Venice, their old collaborator Eero Saarinen launched his groundbreaking plastic "tulip" chair, whose smooth sculpted form was heavily influenced by the Eameses' aesthetic.

The basic thrust of both the Eameses' and Saarinen's work was to mold furniture to the body rather than vice versa. While the Eameses preferred the natural look of plywood, Saarinen fell in love with plastic. Plastic struck Saarinen as the essence of malleability, and thus the perfect material in which to express his daring conception of subordinating material structure to human form.

Saarinen was bitterly disappointed that the plastics industry had not evolved to the point where it could mold the round petal-shaped base out of plastic as well. He had courted disaster and defied gravity by eliminating the legs from the chair. But the base, in order to remain structurally sound, had to be cast in aluminum and coated with a fused-plastic finish to blend in.

"I wanted to make a chair *all of one thing again*," Saarinen would later write with regret. As that *one thing*, he had chosen plastic. "All the great furniture of the past, from Tutankhamen's chair to Thomas Chippendale's have always been a structural whole. . . . I look forward to the day when the plastic industry will advance to the point where the tulip chair will be one material, as it was originally designed."

In 1963, lecturing to an audience of art students, Charles Eames held forth on the subtle differences between working in plastic versus traditional materials.

> Consider a sculptor attacking granite with hand tools. Granite resists such an attack violently: it is a hard material, so hard it is difficult to do something bad in it. It is not easy to do something good, but it is extremely difficult to do something bad.
>
> Plastic is a different matter. In this spineless material it is extraordinarily easy to do something bad—one can do any imaginable variety of bad without half trying. The material itself puts up no resistance, and whatever discipline there is, the artist himself must be strong enough to provide.
>
> I feel about plastic much as the ancient Aztecs felt about hard liquor. They had the drinks. But intoxication in anyone under fifty was punishable by death for they felt that only with age and maturity had a man earned the right to let his spirit go free, to self-expression. Plasticene and the airbrush should be reserved for artists over fifty.

Flubber

Seriously, folks, now that plastic had gone and banished anxiety, concern, and worry from the American home—conquering germs, chores, food spoilage, cigarette burns, and dour blasts from the past—it was free to kick up its heels and go just a little hog-wild. Maybe even a little bit crazy, in a restrained fifties fashion—lampshades on the head, and that sort of thing.

In 1949, in a scene straight out of a Jerry Lewis comedy, GE engineer James Wright awkwardly stood in a corner at a cocktail party in New Haven, experiencing one of those moments of acute self-consciousness common to scientists in social situations. As a tension reliever—and as a bit of a gag—Wright began idly bouncing a piece of plastic putty off the floor as if it were a rubber ball. This tall, Capra-esque figure proceeded to coax a swelling crowd of onlookers over to his corner by calmly stretching the rubber ball into a long taffylike strand, demonstrating its spiritual affinity with chewing gum. For the grand finale of his magic show, Wright took this pink ball of stuff and after pressing it down on a piece of newspaper, peeled it carefully off. He then handed the piece of putty around the awestruck crowd. Lo and behold, it had lifted the printed image off the paper precisely!

One of the guests cracking up at Wright's impromptu pantomime was New Haven store owner Ruth Fallgatter, who happened to be designing a new toy catalog with advertising copywriter Paul Hodgson. After the party, Hodgson and Fallgatter drew Wright aside and asked if it might be possible to include Nutty Putty in their upcoming catalog. After checking with GE, Wright agreed to let them have a small test batch. To Fallgatter's and Hodgson's surprise, their last-minute, amorphous entry outsold every other item in the catalog—by a substantial margin.

Slowly but surely, flourishing in an atmosphere of carefree civilian frivolity, Nutty Putty's elusive potential was made manifest. It had spent seven years in the industrial doghouse, ever since that fateful day in the spring of 1942 when the U.S. War Production Board had sent out a request for synthetic alternatives to I.G. Farben's styrene-based Buna rubber and Standard Oil's petroleum-based butyl. Among the recipients of the request was General Electric, which assigned Jim Wright, one of its top polymer engineers, the sensitive task of coming up with a rubberlike compound for possible use in truck tires.

Wright restricted his investigation to the newly emerging area of silicone plastics, which combined remarkably high melting points with high elasticity. By mixing silicone oil with boric acid, Wright obtained a rubberlike compound that, at first glance at least, looked rather impressive.

It was certainly rubberlike—to the max. If you rolled it into a ball and bounced it on the ground, it rebounded 25 percent higher than a comparably sized crude rubber ball. It was impervious to molds and rot. It easily withstood a wide range of temperatures without degrading. It could be stretched to an enormous length without breaking. It was more elastic by far than crude rubber. But strangest of all, if you pressed a piece of it down on newsprint or a comic book, it "lifted" the image right off the paper. It seemed to be—for what it was worth—inherently photosensitive.

In Wright's opinion, it had potential. The question was, for what? It wasn't strong enough to replace crude rubber in any of the government-mandated applications specified by the War Department. After that determination, the Pentagon brass never had much use for it. As for GE, the only employees who saw anything in it at all were Wright's fellow scientists and lab workers, who couldn't stop playing with "Nutty Putty," as it was affectionately called.

By the end of the war, nobody had figured out any strategic purpose for Nutty Putty. The conventional wisdom around GE was that it was an industrial and commercial dead end. A few corporate higher-ups who fancied themselves cards kept clumps of it hanging around their offices, strictly for laughs. But in late 1945, at Wright's insistence, GE made a last-ditch effort to commercialize the invention. They mailed samples of Nutty Putty to some of the world's leading industrial engineers, challenging them to come up with a practical use for it. Every recipient pronounced himself stumped, though none sent their samples back.

The ultimate use of Nutty Putty, as it turned out, was as a gag, and a novelty. That was all. In an age when fashionable intellectuals professed to admire Sartre, Camus, and Beckett, Nutty Putty epitomized the sublime Absurd.

Ruth Fallgatter, the first American to recognize its commercial potential, inexplicably passed on her chance to usher Nutty Putty into the big time. By default, the field was left wide open to her partner Paul Hodgson, who decided to take a gamble on Wright's bouncy brainchild. He bought a ton of it from GE (laying out the impressive sum of $147) and hired a Yale student to separate it into one-ounce balls and stuff them into brightly colored plastic eggs. Under the name Silly Putty, Jim Wright's synthetic rubber at last achieved the soaring success it deserved. In its first year, Silly Putty sold faster than any other toy in U.S. business history, racking up over $6 million in sales.

The American love affair with Silly Putty was by no means limited to small children. Grown men and women were thrilled to find that the flaky flesh-colored goop did have a utilitarian side. It pulled cat and dog fur off upholstery. It lifted ground-in dirt off woven car seats. Placed under a table or chair leg, it stabilized creaky living-room furniture.

But though the list of household tricks it could play was added to daily, no worthwhile industrial use was ever found for it. That didn't stop Hodgson from retiring from the Silly Putty business a wealthy man, worth at the time of his death well in excess of $140 million.

Self-Cling II

Fall, 1948: Returning from a hard day's hunting in the Alps, the Swiss engineer George de Maestral entered his kitchen with his faithful retriever at his side, covered from head to foot in cockleburs. They clung to de Maestral's coat, they clung to his sweater, they clung to his corduroy trousers. They even clung to the coat of his dog.

While spending the rest of the evening meticulously picking the stubborn hooks out of his dog's hair and his clothes, de Maestral gradually developed a grudging respect for the admirable grit with which those infernal nuisances, formally known as burdock-seed heads, clung to their unwilling hosts.

Taking a few moments to examine the structure of these tenacious seed pods under a microscope, de Maestral readily identified the structural source of the cocklebur's strength: hundreds upon hundreds of tiny hooks, which naturally entangled themselves in the loops of the fabric of his, or his dog's, jacket. Their purpose, of course, was strictly reproductive. The cocklebur efficiently disseminates its own seed by hitching a ride on the furry back of any animal careless enough to rub up against it.

On that crisp autumn evening, George de Maestral decided to see if he could turn one of nature's greatest pains in the rear end into one of the Synthetic Century's greatest miracle tools. He defiantly resolved to duplicate, if possible, the marvelous hook-and-loop system by which the cocklebur propagates itself.

What most impressed de Maestral about the hook-and-loop system was its utter randomness. It didn't matter how precisely the hooks pressed up against the loops, because with hundreds if not thousands of each per square inch, a secure bond was bound to be achieved as long as the two surfaces came into even the briefest of contacts.

It took eight years—until 1957—to perfect a synthetic cocklebur. He called it Velcro, a name derived from the French for "velvet"— *velours*—combined with *crochet*—French for "hook." Laboriously constructing each tape by hand, de Maestral spent the greater part of his time ceaselessly experimenting with a wide variety of materials, both man-made and natural, before settling on nylon as the medium of choice for holding both hooks and loops in place.

Nylon possessed all the desired properties: It could be spun in any thickness required. It didn't break down after repeated fastenings and unfastenings. It didn't rot, mold, or naturally degrade, and it was relatively cheap. De Maestral's greatest obstacle to attaining complete self-cling capability was figuring out some way to craft the hooks. He was stymied, until he realized that if he made loops and cut them in half, the clipped ends would turn into hooks! Velcro was unique among all rival mechanical fasteners—buttons, zippers, snaps, and all other ingenious devices contrived by human beings to keep body and belongings together—by displaying a staggering shear (as opposed to sheer) strength—the force required to pull parallel pieces sideways, in opposite directions.

While ten to fifteen pounds per square inch of force is required to pull two standard Velcro tapes apart—more than needed to split most adhesives—it takes a ton of shear force to pull apart two hundred square inches of Velcro. In an age of tentative commitments, Velcro could be closed and opened hundreds of times without suffering any detectable degradation. The first hook-and-loop fastener was a direct copy of a natural original. Like the geodesic dome, it exploited the inherent properties of a synthetic substance—in this case, nylon—to create a low-tech solution to an age-old problem: the center failing to hold.

Do the Twist

Cut to spring, 1957. Los Angeles. Arthur K. "Spud" Melin of the Wham-O-Manufacturing Company, a well-known novelty firm, receives word at a toy convention of the phenomenal sales of three-foot-diameter bamboo hoops, which have become a school yard craze in Australia.

Throughout Australia, children's calisthenics classes had long used bamboo hoops to give kids precision and balance. Kids down the ages had played with hoops—spliced together from old barrel staves or old iron wheel rims that could be controlled with a hooked iron rod. But these kids in postwar Australia had slyly converted an instrument of authority into an outlet of pure hormonal expression.

Melin and his partner Richard Knerr ordered a batch from Australia and began personally promoting them by handing them out free to neighborhood kids at local parks, playgrounds, and school yards, always taking the time to demonstrate the hoop personally. The basic move involved keeping the hoop revolving around the body without touching it with your hands or letting it drop to the floor. Only vigorous hip swinging did the trick. In the spring of 1957, hooping and hip swiveling could mean only one thing: rock 'n' roll!

Plain old bamboo wasn't sexy or colorful enough for cagey Spud Melin, who called his company Wham-O after the sound of the slingshots they made, which sounded like "Wham-O!" Retiring to his factory in San Gabriel, he had a few prototype hoops turned out of a brand-new high-density, extremely lightweight polyethylene—Phillips 66 Marlex HDPE (high-density polyethylene). Developed in Italy by Guilo Natta, HDPE had just hit the market and had recently become available in a wide range of bright colors. Each hoop cost about fifty cents to produce, and was easily fabricated by bending the superflexible tubing into a circle and then clipping the ends together with a wooden plug and a few metal staples.

Melin called the plastic rings "Hula Hoops," in honor of an ancient Hawaiian dance, the hula, which requires no hoops at all. No matter. The new hoops could be tossed in the air, made to climb stairs, skipped and jumped through while they rolled by, if one were so inclined. But if you followed the accompanying instructions— "Hug the hoop to the backside . . . Push hard with the right hand . . . Now rock, man, rock! Don't twist . . . Swing it! Sway it! You got it!" you were engaging in a primitive form of adolescent sex play, at which, according to *Life*, "Women are more adept than men."

Hula-Hooping gave vent to pagan erotic energies while main-

Courtesy of Mattel Sports

taining a cover of innocent child play (see: Twister). By the spring of '58, as hooping rolled east and even grown-ups began Hula-Hooping to lose weight, hoop sales topped fifteen million, at an average of $1.98 a pop. The Wham-O factory in San Gabriel, California, was churning out twenty thousand hoops a day. Hula-Hooping soon became the largest single use of high-density polyethylene—consuming over a million pounds a day at the craze's peak.

Swim teamers were soon practicing "hoop diving" en masse at local pools. Other kids preferred leaping daringly through fast-rolling hoops without ever touching the sides. An eleven-year-old veteran hoopster from Jackson, Michigan, won a national contest by keeping fourteen hoops in operation simultaneously. Other contortionist hoopsters learned to move separate hoops around necks, arms, and hips at the same time. One Slinky-smooth little boy was featured on national TV for his ability to take his shirt off and put it back on without letting his hoop fall to the floor.

Not everyone was amused. But a few squares tried to prove their cool by trying to see something wholesome and socially redeeming about hooping. One prominent Christian commentator earnestly strove to put a spiritual spin on the disturbingly pagan whirling-dervish mania. "The circle," he solemnly advised, "represents the infinity of God. It is a symbol of Him who has no beginning and no end." The hoop was eternal; the sacred circle. It also represented the ultimate attainment of pointless plastic perfection, circa 1957.

Every craze has its peak and its inevitable decline. By the end of the year, the novelty value of hooping was beginning to wear off for all but the most committed devotees. Once a fad has passed its peak, its more avid practitioners have a way of looking silly. Fortunately, Wham-O had another plastic ace up its sleeve, which was just so much fun to play with that its desirability never died.

Pluto Platter

1957: in the epic year of the Hula Hoop, Wham-O snapped up the rights to an insurance policy for the day hooping died. California inventor Walter Frederick Morrison—who had earned the enduring respect of his peers by inventing a home Popsicle maker—dramatically improved upon an old college fad dating back to the 1870s.

Shortly after the Civil War, William Russell Frisbie opened a bakery in Bridgeport, Connecticut, which became extremely popular with Yale students. Frisbie's pie tins were stamped with the name "Frisbie" right in the middle, which led bored Yalies—who took to sailing discarded pie tins for considerable lengths across the leafy neo-Gothic campus—to cry out "Frisbie!" every time a pie tin went floating by.

Morrison's contribution to the evolution of "the flying disk" was to redesign it in Tennite, a brand of cellulose acetate manufactured by Tennessee Eastman. He not only plasticized the pie tin but dramatically enhanced its aerodynamic potential by molding around its edge the now famous curled-up lip that made it look like a pizza crust, and provided a remarkable degree of stability in flight.

Morrison called his disk "Li'l Abner," after the popular cartoon character. He took to wowing crowds at county fairs around California by tossing it to his wife, who stood a shocking number of yards away, prepared to catch it as it sailed toward her with mesmerizing, pinpoint accuracy. To astonished onlookers, Morrison claimed that the flying disk was actually sliding along an "invisible wire," which he would be willing to part with at a price. The disk itself was scot-free.

Morrison's "Li'l Abner" caught the attention of Spud Melin, just as the Hula Hoop craze was beginning to die down. Melin and Knerr renamed it the Pluto Platter—in an attempt to capitalize on the mounting interest in space travel and exploration touched off by the success of Sputnik—and deftly drove his point home by impressing a planetary pattern into the plastic, adding the now famous flight-stabilizing grooved surface. A year later, Melin decided to rename it the Frisbee, in honor of the tin original.

Lawn Art

1957: a banner year for polyethylene, and a turning point in popular art. In that landmark year, an obscure twenty-one-year-old student at the Worcester Art Museum School, Don Featherstone, received a job offer out of the blue from Union Products, a small novelty-plastics company in Fitchburg, Massachusetts.

Union Products was the nation's leading manufacturer of lawn art, a popular folk-art form that had only recently—since the war—been exploited for its mass-market potential. While the original lawn ornaments were made out of brass or cast iron, and ranged from the still popular lantern-holding lackeys—once rendered exclusively in blackface—to a menagerie of animal stereotypes, by the twenties and thirties concrete lawn ornaments had become popular roadside attractions outside middle-class homes. During the depression, carving and hand-painting flat plywood lawn animals evolved into a cottage industry—a way to make a few extra bucks by sticking them on the front lawn for possible sale to passing motorists. In the space of a few decades, lawn art had evolved from an aristocratic medium evocative of an earlier, feudal time—bronze lackeys holding lanterns—into a popular, even pop medium.

After the war, while flat plywood folk art retained its popularity, Union Products began testing the market for plastic renditions of some of its old favorites, substituting flat plastic foam sheets for the traditional plywood. On the assumption that achieving greater verisimilitude would require retaining the services of a trained sculptor, they sent word to the nearby Worcester Art Museum School that they needed a good artist—willing to work closely with animals.

At first, Featherstone was disinclined to take the job. But after receiving no other offers, and asking himself, "How many rich artists do you know?" and answering "none," he took the plunge and moved to Fitchburg.

Just as Giotto and the perspectivist painters of the Italian Renaissance deepened painting's picture plane from flat medieval iconic images, so Don Featherstone chose to heighten the realism of plastic lawn art by rendering his designs according to life—sculpting wherever possible from live models.

Featherstone's breakthrough design was Charlie the Duck, which followed a less successful 3-D cat, 3-D toadstool, and 3-D fire hydrant—"even the dogs weren't fooled by it," as Featherstone ruefully recalls. Part of the problem was a lack of verisimilitude, but with Charlie the Duck, Featherstone outdid himself—he painstakingly sculpted his clay model from an actual quacker he picked up from a nearby duck farm and kept in his clay room for six months.

Charlie the Duck was a modest commercial success for Union Products. But for Featherstone, it represented a major artistic breakthrough. Now, Don felt confident that Charlie the Duck could be a launching pad to an entire assortment of anatomically correct lawn ornamenture. Now, Don knew for sure that he had what it took to take on lawn art's *Last Supper:* a 3-D pink flamingo.

During the "Dark Ages" of lawn art, anonymous artisans had painted and carved pink flamingos without regard to attaining anything even approximating a unified artistic vision—much less an iconic representation of the full plastic flowering of an enduring fifties aesthetic.

But Don Featherstone knew it was high time to take on the pink flamingo, particularly since Union Products' first tentative contribution to the art form, rendered in flat foam, had been an artistic as well as a commercial disaster. The flamingo was the fifties equivalent of the Madonna-and-Child motif, a popular genre subject for lawn artists dating back to the early forties, when Florida first became a popular vacation spot and flamingo souvenirs—emblazoned on shower curtains, cut into decals, turned into lawn ornaments—became a fad.

Union Products' flat-foam flamingo had an image problem: It looked good from the front, but only from directly in front. From the side, it resembled a billboard. It also deteriorated rapidly in bad weather, and was for some inexplicable reason a preferred snack food for dogs.

Featherstone could not very well get his hands on a live pink flamingo. But he did the next best thing: He found some glossy photos in *National Geographic,* and did his best to render the birds—both male and female—from all angles with unstinting respect for anatomical integrity. Once he had his clay models complete, Don Featherstone had a feeling in his gut that he had a hit on his hands. Little did he know that it was destined to become the masterpiece not only of a lifetime, but of an entire folk genre.

Union Products' expert molders cast an aluminum two-piece mold from Featherstone's clay originals, melted polyethylene plastic resin beads into a fluid state, added a bright pink, fade-resistant dye, injected the molten mixture into aluminum die molds and, after waiting a few seconds for the plastic to assume the shape of the mold, out popped a flamingo—up to two thousand pairs a day.

Unlike inferior competitors (of which there were many) true Don Featherstone originals were always sold in pairs—one female sculpted to stand with its head up, the other, male, with its S-curved neck deeply depressed, as if in mourning. After the flashing was stripped away, the beaks, feathers and eyes were darkened, steel tubular legs inserted, and presto! Two pink flamingos, coming right up! *Flamingo plasticus* hit a home run with the suburban middle class, in Featherstone's opinion, because "a lot of housing looked alike, and people were looking for something to make *their* house look a little different."

Boy, did they succeed!

Twenty-first-Century Synthoid Man

By the tail end of the fifties, America had become thoroughly plasticized. The national polymerization process had occurred without anyone paying close attention, with nary a peep of protest, except for the isolated rumbling and grumblings of a perverse chorus of naysayers—beatniks and naturalists mostly—joined by the occasional mutterings of hidebound aristos harumphing in their wood-paneled dens and clubs that "wood" was no longer "wood" and "leather" no longer leather. Even the funky old woody was sided with "faux wood-grain."

In the good new days, all man-made materials worth their salt sported popular brand names: Dacron and Orlon, Saran and Naugahyde, Tupperware and Wham-O, Silly Putty, Plexiglas, Velcro, Formica. Just plain folks, who had once purchased basic material commodities by generic handles—cowhide, cotton, wool, wood—now demanded products by registered trademark. Plastic's triumph was a triumph of product packaging.

Plasticization was also a triumph of petroleum, from which all the new plastics—once distilled and derived from coal tar—were made. And a triumph of TV, which pushed plastic products with live commercials promising imminent deliverance from drudgery, dirt, and organic disorder. And a triumph of the automobile, which required new, huge, looming plastic signs to communicate to the bypassing masses—giant ice-cream cones, mammoth motel "No Vacancy" plaques, inflatable French fries the size of seals. A triumph of the vulgar and plebeian. And a triumph of the golden age of "Better Living Through Chemistry"—the last time a phrase like that could be uttered without a knowing smirk.

9

Pop Plastic

New York World's Fair 1964–65 Vatican Pavilion—*Pietà*
(Courtesy of The New York Public Library Picture Collection)

Plastic *Pietà*

Robert Moses, the aging czar of New York's 1964 World's Fair, may not have known much about art, but he knew what he liked: Michelangelo's *Pietà*. That was the sort of art plain folks could admire without the benefit of a degree in art history. That was the sort of art the world's greatest World's Fair desperately needed to prove its commitment to celebrating the works of God as well as Mammon.

Through the offices of his good friend Francis Cardinal Spellman, and by the good grace of Pope Paul VI (who had hoped to honor the 400th anniversary of the sculptor's death with some sign of veneration), Mr. Moses prevailed upon the church fathers to permit the six-foot nine-inch white Carrara marble Renaissance masterpiece—all 6,700 pounds of it—to be prized from its pedestal in the dimly lit chapel at Saint Peter's Basilica, where it had lain in translucent serenity for five hundred years, and be hermetically sealed inside a steel crate. After being rolled on dollies from the altar to the chapel entrance, rope-hauled down St. Peter's steps onto the tailgate of a waiting truck, and driven at a maximum speed of twenty miles per hour under armed escort 150 miles south down the Autostrada del Sole to a pier in Naples, the waterproof, airtight floatable crate was hoisted by floating crane onto the deck of the Italian liner *Cristoforo Colombo,* lashed to the ship's deck with bindings designed to release in the unlikely event that the ship became submerged, and outfitted with its own SOS radio beacon.

But even with such extraordinary security precautions firmly in place, "Italians went into trauma over what would be an international tragedy if something went amiss," *Life* gravely reported. Still, once the pope had given his blessing, there could be no turning back. So a consortium of friendly insurance companies stepped in to assume the sensitive role of art guardians, insisting over the strenuous objections of Italian workers that the statue be packed in plastic beads instead of excelsior (fine wooden shavings) to enhance its impact resistance in transit. The Italian art packers, preferring wood to plastic, demonstrated their devotion to Old World tradition by storming out of old St. Peter's Basilica in protest, leaving a team of unsentimental syntho-savvy Americans to complete the delicate task for them.

Even after the *Pietà* made it to New York all in one piece, neither the statue nor its anxious sponsors were close to being out of harm's way. With an estimated fifteen million guests expected to stop by the Vatican Pavilion to pay their respects, it became incumbent upon Broadway set designer Jo Mielziner—*Pal Joey, Annie Get Your Gun, A Streetcar Named Desire*—to conceive of a dignified solution to the pressing problem of crowd control. Mielziner's high-tech design strategy featured an elaborate series of multitiered moving walkways, reminiscent of an airport concourse, guaranteed to provide every visitor with a seventy-five-second window of opportunity in which to gaze upon the work before being mechanically whisked toward the exit.

But whirring walkways alone, while discreetly maintaining a minimum distance of sixteen feet between the crowd and the object of their veneration, could in no way prevent something unfortunate occurring in these troubled times. Additional measures would need to be taken. Measures that might carry the unfortunate risk of compromising the statue's artistic integrity. Either that, or take a chance of sparking an international incident by losing the work entirely.

A few days before the fair's official opening in April 1964, the Vatican hosted a solemn invocation to celebrate the illumination of the *Pietà* in its new context. After Cardinal Marella, the papal legate, read the pope's blessing, the sculpture was starkly illuminated by a cold blast of flickering blue light. As forty-three high princes of the church, gorgeously adorned in traditional regalia, led by Cardinal Spellman hoisting aloft his golden crosier, marched in joyful procession around the pavilion rotunda, a few of the more theatrically minded among the luminaries in attendance burst into a spontaneous round of applause. This profane acclaim lasted just a few seconds before, as the *New York Times* sternly commented the next morning, "those who applauded recognized that applause was inappropriate, and quickly stopped."

Inappropriate or not, applause had been the anticipated response to the distinguished Jo Mielziner's elaborately conceived, expensively executed set. But no sooner had the clapping quieted down in Queens than the Bronx cheers began blowing in earnest.

"When they finished installing the *Pietà,* the Fair sat back and waited for applause," *Life* explained. "But all it got was loud boos." Not to mention brickbats. The *Times* gave a pale hint of the troubles to come in a revealing front-page subheadline: "Plastic Shield Stirs Debate over Setting."

"It became apparent almost immediately," the paper of record elaborated, "that the setting might prove controversial. . . . The most pointed criticisms were directed at the transparent plastic shield that had been placed before the sculpture to protect it." Some of which the *Times* saw fit to print in a barbed piece by art critic John Canaday, who pronounced himself capable of feeling "nothing" when gazing upon the *Pietà*'s garish setting "but a sense of violation."

Bad enough that those "distracting blue lights . . . blinked with a terrible automatic impersonality." Mielziner had seen them as subtly evocative of the solemn guttering of votive candles. Any symbolic association with the Virgin Mary in no way alleviated, in Canaday's opinion, the pronounced "edginess of the setting." Even worse was the corporate decision to seal the statue inside a "transparent vacuum . . . where the statue looks somehow helpless and cold, as if being subjected to refrigeration."

By "transparent vacuum" Mr. Canaday was clearly referring to the bulletproof Plexiglas shield prudently put into place to protect the *Pietà* not only from bullets and bombs but—as one nail-biting official confidentially put it—"some nut who might throw a bottle of ink." The designers had regarded the floor-to-ceiling curtain, in its serene transparency, as a relatively innocuous presence. But they had gravely misjudged their audience, not to mention their subject. Nothing, not even the faux Gregorian chant background Muzak or the showy overhead cross, could have been more compromising to the lofty spirit of the Renaissance than that hideous plastic window, which miraculously succeeded in degrading an original work of art into a cheap replica of itself.

But the most horrifying association evoked by that Plexiglas shield could not be blamed on Mielziner or Moses alone. By a tragic twist of fate, the '64 World's Fair opened just five months after the assassination of that other great Catholic monument, President John F. Kennedy. When faced with the grievous aesthetic assault of the plastic shield, few onlookers could forget that the late president's final executive decision had been to reject the offer of just such a bulletproof shield for his open-roofed car in the Dallas motorcade. Five months later, by "sealing the statue behind a transparent vacuum," the powers-that-be had unwittingly touched a raw nerve. "Yes, Virginia," that shield seemed to say, "it is a random, violent world out there, chock-full of nuts just dying to shoot bullets and hurl bombs at a much beloved icon."

The plastic shield surrounding the *Pietà,* widely interpreted as a sad commentary on the growing sickness of American society, was swiftly enshrined as a symbol of not only all that was wrong with the World's Fair but much that was wrong with an increasingly plastic world. To be scrupulously fair to Moses and his detractors, it couldn't have been easy feeling good about the future while locked in a present marked by the violent death of an American president, the murder of three innocent civil-rights workers in Mississippi, the assassination of Malcolm X, and the imminent onset of saturation bombing in North Vietnam.

The Beatles' "I Want to Hold Your Hand" may have been the number one pop song in 1964, but a new note of cynicism had hit the mainstream big-time. That was the year that the Nobel Prize for literature was awarded to Jean-Paul Sartre, that John le Carré's dark *The Spy Who Came in from the Cold* topped best-seller lists, that the old blacklisting Hollywood finally came clean on the cold war with Stanley Kubrick's sardonic black comedy, *Dr. Strangelove.* It was a time of widespread disenchantment, not enchantment, with naive and sentimental notions of global progress.

These days, one couldn't even open a world's fair without provoking civil unrest. On opening day, Moses himself was forced to put down a demonstration on his home turf, as civil-rights activists vowed to blockade every road, bridge, and tunnel leading out to the fairgrounds to protest the stalling of President Johnson's civil-rights bill in the Senate by a cadre of Dixiecrat senators.

No matter how much Mr. Moses and his minions sought to stir up the fun, the vision of the future he and his corporate cronies had conceived by committee was simply too bombastic, sterile, and grandiose for many visitors to accept at face value. Exhibits that managed to avoid being altogether in bad taste often came off as simply absurd—sometimes sublimely so.

American Express's dubious decision to laminate multimillions in megabucks—real-world currency—into clear plastic "leaves" on a steel-cage-encased Money Tree had been clearly intended as a witty commentary on the old saw "Money doesn't grow on trees." But in light of the growing awareness of pervasive poverty not only in foreign countries but in our own backyard, the joke fell flat on its embossed plastic face. Typical of criticisms leveled at the Money Tree was the offhand observation that "the large amount of cash laminated into uselessness on the Money Tree provides the best symbolism of the 1964/65 Fair."

In a *New York Times* special issue devoted to the fair, Mr. Moses grandly predicted that, after the *Pietà*, the fair's number one "cynosure—what plain folks call a sight for sore eyes" would be Walt Disney's animated statue of Abraham Lincoln, the oversized star of the Illinois Exhibition. But more onlookers were appalled than inspired by this grotesque exercise in Walt Disney's ersatz science of "Animatronics," a Paul Bunyan–sized android reciting the Gettysburg Address in a dull monotone reminiscent of an imprisoned soldier's repeated incantation of name, rank, and serial number. Robert Moses proudly proclaimed, "The stars of my show are Michelangelo and Walt Disney." But cultural connoisseur Richard Schickel could scarcely refrain from smirking: "Are we really supposed to revere this contraption, this weird agglomeration of wires and plastic?"

To Yale architecture professor Vincent Scully Jr., the answer was simple: "Walt Disney's Lincoln caters to the kind of phony reality . . . that we all readily accept in place of the true." The polarity of responses from reverence to disgust at the plasticized *Pietà* and the mechanized Lincoln revealed a widening fault line in American cultural politics. "Mr. Disney, I'm afraid," Mr. Scully insisted, "has our number. But so does Mr. Moses; for when all is said and done, his Fair is exactly the kind of world we are building all over the U.S. right now . . . nothing but the concentrated essence of motel, gas station, shopping center and suburb."

Here, at last, was plastic reality—a naturalist's nightmare come to pass atop the filled-in salt marshes of Queens. The most depressing thing about this brave new synthetic universe was that everything in it seemed to be set off by quotation marks: trees had become "trees," and leaves "leaves." Nothing was what it was, but some spiritually impoverished simulacrum of what it was supposed to be.

"The Fair will offer the materials supplier and processor a combination showcase-laboratory facility in which to test the application of plastics in building, construction, transportation, communications, sign displays, packaging, and lighting," crowed *Modern Plastics* magazine in a special issue devoted to the fair. But what would it offer real people? Gazing in horror at the concrete-bound "City of Tomorrow" envisioned by the Pentagon-style planners ensconced deep within the bowels of General Motors, one saddened observer saw nothing but a vast wasteland worthy of T. S. Eliot. As one critic carped, he found "no parks, no grass, no trees . . . a city engineered for machines but not designed for people, not for me."

Another dismal symbol of the fair's blissful neglect of new ideas and fresh visions were the unspeakably horrid plastic-topped "trees"—fashioned out of tubular steel and resembling hat racks—erected to provide "shade" to footsore crowds. But this was plastic "shade," cleansed of all harmful infrared rays, sucked dry of heat, provided by thousands of identical plastic gray-tinted acrylic "canopies" promoted by their manufacturer as "ideal for backyard or patio." One of the most highly touted features of these so-called trees was that they were capable of funneling rainwater into their "trunks." But in keeping with the corporate retro-vision of the fair, this breakthrough design concept had made its debut at the 1959 *Moscow* World's Fair. That the powers-that-be lording it over the cultural capital of the world's most powerful country, a land perpetually locked in philosophical and moral battle with a far cruder enemy, had seen fit to provide an international technological showcase with a five-year-old notion conceived by central planners in the Soviet Union was yet another sign that something was desperately amiss in the state of New York, and in the state of the nation at large.

World's fairs, by design, are temporary affairs. So it came as no great surprise that many pavilions were extruded, heat-stamped, and blow-molded from a stunning variety of plastics and other synthetic materials. But what irked so many of the fair's "constantly carping critics"—to quote Robert Moses—was its sedulous stress on the surpassing wonders of artifice and progress. The very notion of "progress" seemed increasingly frayed by reality—a dubious abstraction set off by quotation marks. Who *but* Robert Moses and his cadre of cultural conservatives still took progress seriously? Didn't *they* know President Kennedy was dead?

Even the GM Futurama, which in 1939 had wowed the most cynical visitors with Norman Bel Geddes's sleek, whizbang "City of Tomorrow," insisted on carrying Prince Albert's century-old grand design of "conquering nature to man's use" to alarming extremes. The future as seen in retrograde vision by GM consisted of a series of nightmarish fantasy landscapes dominated by mammoth machines, where ordinary people seemed at best an afterthought, at worst a bacterial infection.

The star of the Futurama was the Self-Propelled Road Builder, five stories tall, three football fields long, a vast automaton capable of gobbling acre upon acre of tropical rain forest by hewing down centuries-old tree trunks with foremounted laser beams—this fair clearly had it in for trees—as it lumbered forward on caterpillar tracks, digesting jungle while simultaneously exuding a sleek black-top ribbon of highway, four lanes wide, in its wake.

Over at the Festival of Gas, a generic utility exhibition, the Self-Propelled Road Builder had a domestic equivalent: the Norge Dish-maker, a sleek, dishwasher-sized device equipped with a sliding transparent plastic window that gave it the look of a portable Automat. This tidy little critter could chew up an evening's supply of dirty plastic dishes, plates, cups, and saucers, grind the residue into plastic pellets, melt those pellets and roll the resulting molten mass into styrene sheets. These were then vacuformed into the next day's supply of new plastic plates, cups, dishes, and saucers overnight—all while the family relaxed 'round the TV.

Absurd, perhaps. But at least the Norge Dishmaker looked for-ward to a future of less than limitless resources, in which plastic recycling might one day play a key role in a strategy of planetary preservation. Its manufacturers, lacking the confidence that envi-ronmental concerns would ever sell appliances, preferred to tout the Dishmaker's capacity to "conserve space in the kitchen."

For die-hard plastic fans, there was still much to be enthralled by at the fair. The Sinclair Oil exhibit—another Disney-designed Animatronic extravaganza—offered a life-sized T. Rex devouring a three-foot-long plastic orange, while kids were inspired to vacuform their own miniature dinos in a glass-sided contraption where they could watch the petro-product plastic vividly flow into a bifurcated mold. Presto! a little green dino popped warm as a fresh loaf of bread into your palm. If nothing else, the self-operated bronto-maker sent a clear message about where today's plastic was coming from: the world's fossil-fuel reserves.

Formica House, artfully constructed inside and out of that "wipe-clean wonder," was praised by no less an authority than *Good Housekeeping* as "a house without strain on eye, ear, temper or muscle." But by the mid-sixties, the idiotproof "Formica idea" had so thoroughly saturated the American psyche that the very notion that "every eye-delighting surface can be wiped clean in seconds" was no longer a distant dream but a quotidian reality. Pleasant enough, but certainly nothing to write home about. And certainly not worth placing a video call on the Bell Picture Phone being trumpeted over at the Ma Bell pavilion—another high-tech Frankenstein seriously out of sync with its times.

At Du Pont's "Live Theater" Pavilion, a faux Broadway musical called *Wonderful World of Chemistry* (with high-tech costumes by Oleg Cassini) remained in continuous repertory, delighting SRO audiences with such sprightly numbers as "The Happy Plastics Family." But the theater itself—walls, roof, doors, floors—was itself all part of a synthetic "environment" where the transcendent theme of artificiality had been carried to predictably ludicrous lengths. From the Tedlar roof to Delrin door knobs to Antron and Fabrilite seat fabrics, Dacron fire hoses, Zytel hinges, Mylar stage curtains, Hylene polyurethane foam undercarpeting, the entire space was a spun-woven example of "Better Living Through Chemistry."

Apart from routine razzmatazz, the most striking thing about the brand names on display was how faintly their names resonated with a larger public. Zytel and Delrin and Antron and Fabrilite were all admirable products, no doubt. But not exactly household words, like nylon or cellophane.

In a series of press releases, Du Pont had dangled the prospect of a new release that would wow the world, a worthy synthetic successor to nylon. In order to assume its rightful place in the top tier of the chemical pantheon, this new hero would have to slay its own dragon—by bravely displacing some venerable natural material, scattering the forces of nature before it with the greatest of ease.

But when the time came to show its cards, the target turned out to be shoe leather; the conquering hero was something called Corfam. As journalist Peter Lyons grimly noted in *Holiday*, "To glamorize such a product strains to the utmost the resources of show business."

Du Pont's Edsel

From a strictly commercial standpoint, market analysis had revealed that—in the words of one respected in-house authority— "Leather seemed ripe for substitution." In the coming decades, human population was expected to outstrip cattle production, leading to acute leather shortages. Apart from the nice prospect of price boosts due to unmet demand, leather exhibited a number of more intrinsic shortcomings: Irregularly shaped hides made large pieces awkward to cut; even the most carefully tanned hides stiffened after being wet and dried; and the bane of all natural materials, high maintenance requirements.

In short: Leather grew old, with or without trousers rolled. Over time, leather rotted. Leather (sometimes) stank. It had to be scraped clean and tanned, tainted, and dyed. Leather could be said to possess all the generic defects of its class: It wasn't rough enough or tough enough to take it in the high-stress modern world.

But in those vices, *Fortune* sagely reminded the syntho-wizards at Du Pont, lay venerable virtues: Leather grew more supple and tender with age. Natural hide was a tough act to follow because it was "at once tough and pliable, soft and hard, formable to shape yet resilient, dense yet porous to air and moisture, and available in a large variety of finishes, textures, and surface patterns." Not to forget that "around this material . . . clings an aura of romance stretching back to the dawn of history. It still moves some men to prize, and caress fine leathers, and to reject with horror any chemically devised alternatives."

Leather had been a prime target for Du Pont's streamlined synthetic army ever since the years just before World War I, when the old Du Pont Fabrikoid Company first began cranking out pyroxylin-coated fabrics called "leatherettes." While they and their synthetic successor, "split vinyl," did a passable job replacing leather in automobile upholstery and in some medium-priced luggage, neither could be considered a prime contender to displace top-grade leather in the fine shoe-upper market.

As *Fortune* observed, the synthetics "had little of leather's combined flexing strength, shape retention, and lateral give." Or, just as critically, its peerless ability to "breathe or to absorb and dissipate foot perspiration." Breathing was the key obstacle to progress, a direct result of skin's complex molecular structure. But by the mid-fifties, having invested over two hundred man-years of effort in the attempt, Du Pont's fabled Fibers, Film, Fabrics and Finishes Division—known to its familiars as "F & F"—felt sure it had come up with something astonishingly akin to the real McCoy.

Structurally, Corfam was nearly as complex as leather itself: a three-ply laminate composed of a layer of polyurethane foam designed to faithfully mimic the soft, fibrous woven mat of collagen that gives leather its extraordinary strength and suppleness. As far as "breathing" went, Du Pont had developed a unique outer skin of expanded vinyl pierced by a million microscopic breathing holes per square inch. The clever wordsmiths in Du Pont's marketing wing had even coined a new term for this new synthetic, "poromeric," merging "polymer" with "porous" to signify the first porous polymer.

The American Podiatry Association, a leading authority on footwear comfort, came in on the side of the synthetic angels well before the kickoff when it officially announced (after conducting an extensive series of Du Pont–sponsored field tests) that "poromeric shoes appear to be comfortable and compatible with good foot health." By early 1962, F & F was confident enough of Corfam's leatherlikeness that it initiated a larger round of tests.

Recruited into the secret pool were the foot patrolmen of the Newburgh, New York, police department—site of Corfam's pilot plant. Cops reported favorably on Corfam's capacity "to be dunked knee-deep in muck and then simply put under a running faucet or wiped clean with a wet sponge." As for comfort, preliminary results were promising. Corfam's vaunted superiority over leather was confirmed in a number of key areas: weight—up to a third less than its natural counterpart—moistureproofing, and of course, ease of maintenance.

By 1963, thirty high-quality shoe manufacturers, including such chic names in women's footwear as I. Miller and Palizzio, and venerable men's bootmakers Johnston & Murphy, French, Shriner & Urner, and Bostonian, were ready and eager to enlist in Du Pont's campaign to make leather go the way of the dodo. The lone holdout was hidebound Florsheim, which publicly proclaimed its undying loyalty to leather even in the face of Corfam's astonishingly successful debut: three hundred thousand pairs snapped up in the first two weeks, with reorders heavy.

As with nylon, Du Pont deliberately set Corfam's base price at a level equivalent to top-grade cowhide, so no one would get the idea that Corfam shoes were in any way "cheap." After the launch, preliminary press reports consistently claimed that Corfam customers were impressed by all three faux finishes: smooth, grainy, and suedelike. Even "suede" Corfam could be "easily wiped clean with a damp cloth," providing that cloth contained a dash of soap.

Buoyed by this overwhelmingly positive market response, Du Pont confidently cranked up its sparkling new production facility in Old Hickory, Tennessee (a Nashville suburb), and hastily expanded its pilot plant in Newburgh, New York. In its first year, over seventy-five million pairs of "poromeric" shoes were sold at premium prices—an undeniably impressive debut. All signals were go; Du Pont was poised for liftoff.

But not so fast, guys—the incumbent leather industry refused to lie down and die. Unlike silk, leather did not need to be spun by worms in the remote corners of Asia, but was a homegrown, farm-bred natural commodity, as American as Mom, apple pie, and chicken-fried steak. No matter how slickly Du Pont aimed its sights at it, leather would not go away, because every time a red-blooded American tucked into a juicy steak—or lamb chops, or pork—a piece of leather had been released into the market, just waiting to be turned into a bag or shoe.

What ended up killing Corfam was not price nor promotion, but rising consumer resistance. Folks just didn't like it, and once they wore it, they forswore it: Some 15 percent of customers complained of hot, sweaty feet—more than twice the negative-response rate reported by the first round of testers. Though Du Pont officials shrilly insisted that they could "prove scientifically that Corfam breathes better than leather," the very need to mount a "scientific" defense was a sure sign that Corfam was hurting.

By decade's end, Corfam had unofficially earned the dubious title around the corridors in Wilmington of "Du Pont's Edsel." With losses piling past the $70 million mark and climbing, comparisons to the Edsel were painfully instructive. Like Corfam, the Edsel had fallen victim to a research campaign that failed to detect the depth of resistance to, among other design quirks, a disturbingly vulvalike vertical front grill. In 1970, Du Pont quietly sold off all its remaining Corfam product and plant to the People's Republic of Poland, where (perhaps predictably) it prospered.

Pop Art

If there was one thing Robert Moses hated more than Modern Art it was people who liked it—or claimed to. Between Moses and modern-art lover New York governor Nelson Rockefeller, there was very little love lost, not since the governor had taken it into his head to systematically dislodge the curmudgeon from his cluster of powerful posts in state government by pawning him off on the World's Fair Commission as its president, his last post before retirement.

When a ragtag coalition of artists' groups beseeched Moses to consider reconstructing the old Contemporary Arts Building, a welcome beacon for modern art at the 1939 fair, the group's proposal to erect a permanent structure—of aluminum spandrels and rubberized plastic panels—didn't fly with Moses. This was, after all, the man who in his capacity as parks commissioner had ordered the Contemporary Arts Building demolished twenty-five years earlier, willfully ignoring the fact that it had been offered— free of charge—to the borough of Queens to serve as a permanent modern art center.

"They never had any money," Moses dismissively scoffed to a *Times* reporter pressing him for the details behind his decision. "Certainly not that was ever certified to by any bank, corporation, institution or substantial, reliable financial entity." In Moses's world, money talked; nobody walked. "Culture," Moses had flatly decreed before declaring the subject closed, "is somehow associated with long walks and aching feet."

As a result of this and other caustic remarks on the complex relationship of art and society, *Art News* magazine dubbed Moses "the Art Slayer." In the face of such stubborn resistance, Governor Rockefeller—a founding patron of New York's Museum of Modern Art and a latter-day Medici to a raft of contemporary artists— decided to take a few aesthetic matters into his own hands. His sole ace in the hole was the New York State Pavilion, to be designed by his friend Philip Johnson, who gleefully swung into the spirit of the occasion by playfully reinterpreting a rejected proposal for a giant circus-tent-like structure submitted by Frank Lloyd Wright to a Chicago World's Fair committee in 1933.

Wright's great plastic tent was transformed by Johnson into an austere curtain-walled space topped by a whimsical cable-hung roof, providing the perfect circus-tent atmosphere for an exhibition composed mainly of pop art.

When it came time to hang the show, however, the scene beneath the big top quickly degenerated into a media circus. The first controversy—and an omen of more *Sturm und Drang* to come—was sparked by Andy Warhol's mural *Thirteen Most Wanted,* a collection of silk-screened mug shots of fugitives "most wanted" by the FBI. Not only had it been intended—like Robert Rauschenberg's equally controversial mural *Skyway* (which transformed the late president into a Monroe-esque pop icon)—to send an ominous message on the subject of violence in American society, but it fairly resonated with homoerotic overtones.

The governor didn't give a hoot about homoerotic images, latent or blatant. Or antiviolence protest messages. What got his goat was Andy's temerity—or stupidity—in selecting a cluster of certified "mafiosi," some of whom had been acquitted at trial *after* their capture by the FBI. Concerned that a display of such "innocent" mugs might lead to massive civil suits being lodged against the state, Rockefeller respectfully asked Warhol to paint over the offending panels. Which, albeit grudgingly, Warhol did. But only after suggesting to an aghast Philip Johnson that he intended to replace them with a heroic silk-screen portrait of Robert Moses, to be rendered in suitably Stalinist style.

At the very last moment, Johnson succeeded in heading Andy off at the pass. After pleading with the artist that "it would be in bad taste . . . to thumb our noses at authority," Johnson got Warhol to cave. What Johnson had failed to grasp was that thumbing your nose at authority ranked up there with bad taste as one of the cardinal tenets of the pop movement.

But Warhol's backhanded tribute to Moses was, like much of his best work, both less and more than it seemed. Though it irked Moses to no end, the old man was destined to be hoisted on a pedestal by his worst enemies as—God forbid—an unwitting champion of pop art. While the highbrow critics at the *Times* were busy bashing Moses for his blatant commercialism, and *Art News* was moaning that his "taste . . . combines the tone of carny shrill with the spirit of a black-marketeer," a prominent pop critic was fulsomely praising his "pairing of the *Pietà* and [Walt Disney's] 'It's A Small World'" as "the very essence of Pop sensibility." Seen from a pop perspective, even the plastic shield surrounding the *Pietà* conveyed "a subtle sense of incongruous irony similar to that which had inspired Marcel Duchamp's 'Mona Lisa With a Mustache.'"

Meet Jo Mielziner: no *schlock* artist, but unconscious *pop* artist.

"I am for Kool-art," chanted Claes Oldenburg in his famed pop manifesto, "7-Up art, Pepsi-art, Sunshine art, 39 cents art, Vatronol art, Dro-bomb art, Vam art, Menthol art, L & M art, Ex-Lax art, Venida art, Heaven Hill art, Pamryl Art, San-O-Med art, Rx art, $9.99 art." So, for better or worse, was Robert Moses.

It delighted pop artists that sixteen out of forty-four objects representing our civilization to the human race five thousand years hence, solemnly selected by a former curator of the Smithsonian Institution—slated to be slammed into subterranean orbit beneath the soft soil of Flushing Meadow beside the first Westinghouse Time Capsule buried back in 1939—were plastic or synthetic.

Nylon bikini (stars-and-stripes pattern)
Plastic wrap
Electric toothbrush
Ballpoint pen
Reels of microfilm
Credit cards
Beatles record
Freeze-dried foods (encased in plastic)
Synthetic fibers
Heat shield from *Aurora 7* (silicone-and-composite material)
Film history of the USS *Nautilus*
Film identity badge
Fiber-reinforced material
Pocket radiation monitor
Birth control pills (plastic memory-enhanced case)
Plastic heart valve

Gaze upon them, ye mighty, a latter-day Ozymandias might have warned, and be: (1) amused (2) despairing (3) impressed (4) appalled (5) all of the above.

Plastic Man

YOU GET MORE WITH LES!

You are now sitting in New York's finest Canadian Restaurant, the autobiographical culinary environment of Les Levine, one of America's foremost artists. Mr. Levine, born and raised in Ireland, emigrated to Canada at the age of 17. . . . After many years in Canada, he moved to New York City where he now resides. We welcome you to Levine's Restaurant, where Mr. Levine shares with the New York Community a taste of his childhood memories.*

*SPECIAL 20% DISCOUNT IF YOU ARE A LEVINE

Levine's Restaurant, offering "Irish-Jewish-Canadian Cuisine" to New York's art world, was the brainchild of Max Ruskin, owner of Max's Kansas City, the preeminent sixties artist hangout and nocturnal nerve center for Andy Warhol and his crowd.

On opening night (Saint Patrick's Day, 1968) Les Levine, who habitually dined every night in front of a TV set on TV dinners, was forced to eat alone in the nearly deserted back room of his own restaurant, scowling over a bowl of lamb stew and matzo ball soup while one of Ruskin's trusted associates slyly needled him about his capacity to keep "the art world away in force."

"You think people don't like you," the aide softly sneered. "Well, you're wrong, they *hate* you!"

To which Levine replied with a philosophical shrug,

"You're nobody until somebody hates you."

Even Andy Warhol, who typically couldn't be bothered to hate anyone, found Les Levine's work "unacceptable."

"Why it's just *nothing*," Warhol stammered, "I mean, it's just plain *plastic*."

What irritated his fellow artists about Les Levine was that since moving to New York from Toronto in 1964, he had enjoyed a phenomenal burst of popularity based on a canny exploitation of . . . plasticity. His fleeting fame—Warholesque in its brevity, yet

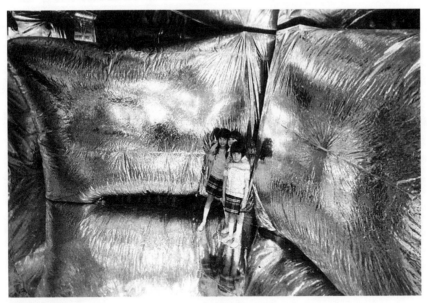

Courtesy of Les Levine

ferociously intense while it lasted—included a splashy profile in
Life:

> One of the glossiest reputations in the art world currently
> belongs to Les Levine, a New York artist who can't draw and
> has never used a conventional art material in his adult life. For
> the past three years, he has been provoking the art scene with
> an alarming array of vexing objects and deeds. His specialties
> are "disposable" art and environmental "places," which may
> consist of nothing more than a series of walk-through plastic
> bubbles or a floorful of 31,000 photographs mired in colored
> gelatin. Levine, who's familiarly known as Plastic Man,
> belongs to no school or movement; he is so ubiquitous that he
> can qualify as a one-man movement. . . .

It went without saying that Levine belonged to no school or
movement because no self-respecting school or movement would
have him. "Somehow, the art world won't accept me," Levine candidly complained to his friend David Bourdon, one of a handful of
art-scene regulars willing to accept him on his own fluid terms.
Though Levine was convinced that the Manhattan cold shoulder

was a result of his own distinctive ethnic makeup—Irish-Jewish-Canadian—Bourdon preferred to ascribe the chill in the air to Levine's acerbic personality. "He speaks of his audience in clinical terms," Bourdon confided in a profile of Levine, "comparing them to 'organisms' that need to be 'collated with the environment.'"

One gallery owner of Les's acquaintance had a simpler explanation: "Les Levine is too smart to be an artist." But one thing about which Levine and his chorus of critics could agree: He managed to be one of the most prolific artists on the New York scene without having to lift—much less lend—a hand. Levine preferred working with the phone. Like an aesthetic executive, Levine had the ideas and phoned in his orders to the plastic pros. For one thing, it was safer.

In an era of rampant mass production, it struck Levine as absurd and romantic to conform to the obsolete requirement that true art should consist only of objects composed of traditional materials worked by traditional means: that is, manually. With regard to plastic, Levine had learned early on that the stuff can be toxic and is most safely fabricated under factory conditions, preferably in a well-ventilated space properly equipped with industrial-strength evacuation equipment. Either that, or you risked doing yourself, your lungs, and your kidneys serious long-term damage.

What Levine preferred, when not whiling away long hours in a studio, was to run around town in a constant state of alert, refining his ideas to a point that the *idea* became the end, not the art object. In a self-conscious parody of capitalist methods of mass production and distribution, Levine mass-manufactured a line of so-called disposables from expandable polystyrene—better known as Styrofoam—which emerged straight from the factory in sixty different styles and thirty colors. Levine proceeded to distribute them in bulk packages of ten thousand units to buyers, retailing at a volume price of $3 per unit.

That Levine scanned the art world with cool cash-register eyes was of course all part of his act. But Levine took the renegade role still further by heretically claiming that he didn't care *what the objects looked like*. The point was for people to "use" his art for as long as they felt like, and when they grew tired of it, to chuck it.

"To be disposable," Levine proclaimed in Oscar Wilde–like epigrammatic style to Rita Reif of the *New York Times*, "something must be made to be destroyed as soon as the owner wishes."

Touché! "To be truly disposable," he continued, "a work of art must be as available as Kleenex, and cheap enough to throw away without compunction." Because "in a fantasy-oriented, consumer-oriented society, culture is just one more thing to be consumed."

When Levine blew up a 1,600-square-foot sea of clear plastic bubbles—Levine termed the "Star Garden" an "architectural device"—on an outdoor terrace at the Museum of Modern Art, he insisted upon informing the SRO crowd in attendance that "the structure is made of clear Acrylite plastic sheets which have been heated and then shaped by jets of air into rounded forms." Otherwise known as "blow-molding." And that the acrylic sheets had been donated by American Cyanamid.

Once the tent was struck, Levine had a problem. "Want to Buy a Sea of Bubbles?" the headlines blared from coast to coast, since the "quirky" story had been picked up—courtesy of Levine's private publicist—by the AP news wire. The Walker Art Center had agreed to ship the bubbles by truck to Minneapolis for its next exhibition, but after that the bubbles had no place to go, no proper home.

But wasn't that the point about plastic? After you made something out of it and you got tired of it and wanted to get rid of it, where in hell were you supposed to put it? It was so easy to make, and so devilishly hard to get rid of.

One man who had since come around to the view that Levine's way of thinking might be worth emulating was his old nemesis Andy Warhol. Now, after the MOMA show, Les Levine's work was no longer "unacceptable." Andy had an idea—a dim one to be sure—that a radically reduced version of *Star Garden* called *Star Machine* might be *just the thing* for the business office of his new factory on Union Square. He tendered a preliminary offer of $5,000 through their mutual friend David Bourdon. But Levine, belying his oft-repeated refusal to be "possessed by material objects," decided that he would rather trade than sell—one Warhol for one Levine.

To discuss the possibility of doing a deal of some kind, Bourdon took Warhol by Levine's loft on the Bowery. "The two men sat soundlessly at a round table flanked by two television sets," Bourdon later recalled, "which ran soundlessly throughout the meeting. Each of them was about as vivacious as a computerized memory bank. From time to time, they covertly scanned each other in sweeping glances that always returned to rest on one of the TV screens."

To break the ice, Bourdon awkwardly advised Levine: "Andy wants to know where to buy cheap plastic."

To which Warhol protested, "No I don't," before sheepishly conceding: "I have to get . . . I thought of doing boxes in plastic. Where can I take a box and get it . . . uh . . . made in plastic or put in plastic or something?"

"You can get it at a place called Just Plastics," Levine briskly replied.

"How come you don't have a plastic tablecloth?" Warhol baited his host, trying in his stilted style to keep the conversational ball rolling.

"I've got a plastic suit," Levine replied, referring to the white vinyl suit he wore on special occasions in his public role as Plastic Man. "But the reason I don't have a plastic tablecloth," Levine shyly volunteered, "is that I really want to move out."

"Find anything yet?" Andy perked up, ever eager to discuss real estate.

Levine allowed that he would probably remain on the Bowery.

"You know, the funny thing about plastic," Warhol free-associated, "is that it's really expensive. You know, they tell you, you know . . . get plastic bags, it looks so cheap—but it's so *expensive*. And when I went out to buy a plastic box the other day, it was a dollar fifty. I mean, a dollar *fifty* for a plastic *box* . . . and it was a very cheap plastic box, not the pretty kind." They never did do a deal.

In June, Warhol was shot by Valerie Solanas, an obsessed member of his own entourage. Not long afterward, he telephoned David Bourdon from his hospital bed.

"I wish it had worked out," Andy sighed softly. "I really wanted those two pieces of plastic."

"You mean *Star Machine*?" Bourdon asked, shocked by the intensity of his regret.

"If I had had it," Warhol moaned, "I wouldn't have been shot. I could have hid in between."

Future Fashion

When Zandra Rhodes, the daughter of a fashion instructor at Medway College of Art, graduated from London's Royal College of Art in 1964, she and a fellow unemployed art-school graduate, Alex MacIntyre, set up house in a dingy flat in St. Stephen's Gardens. They couldn't afford furniture, but they could afford plastic. So they set about, Zandra would later wistfully recall, "creating our own Pop environment, a perfect world of plastic, true to itself, honestly artificial."

> We covered everything in plastic by the yard, including the walls and even the television set. We had plastic grass carpets, collected plastic flowers and trees, used synthetic marble and Fablon tiles. We made the furniture by drawing shapes on the floor and building laminate-and-foam seating and tables on those spots. We built standard lamps from pillars of plastic, and circled them with neon bulbs.

She may have "looked like a mixture between Carmen Miranda and Gina Lollobrigida," according to *Women's Wear Daily,* "with big earrings and platform shoes and lots of make up." But Rhodes and her friend were strictly conforming to a new British pop sensibility evolving in London under the demonic tutelage of journalist Lawrence Alloway, architectural critic Reyner Banham, and the artists Eduardo Paolozzi and Richard Hamilton.

The original pop circle were all members of a crowd that called itself, for obvious reasons, the Independent Group. It had splintered from the London Institute of Contemporary Art, which under the directorship of Herbert Read (the self-appointed evangelist for Modernism in Britain) served as the British outpost for Continental High Modernism—the stark religion of Mies and Le Corbusier.

In 1956, a momentous event occurred across the pond: The new Cadillac Fleetwood rolled off the line in Detroit, flaunting great whopping tail fins. Read, a staunch advocate of the austere "less is more" tradition, solemnly asserted that these grotesque impediments were nothing but the latest and most egregious example of what high-brow designers dismissively derided as mere "Borax"—that distinctly American penchant for elevating form over function, more vulgarly known as "styling the goods."

But Reyner Banham, one of Read's protégés, adored fat fifties American automobiles, with their bomberlike wraparound windshields, bullet-nosed bumpers, tail fins, and air scoops, for precisely the same reasons his mentor Read despised them: They were grandly baroque in the vernacular tradition. It was a case of Popular taste doing an end run around Elite taste.

The schism between Banham and Read divided along generational lines, with Banham and his young crew decamping from the ICA in high dudgeon and setting up their own rival art institute. To distinguish themselves from the suddenly staid Institute of Contemporary Art, they called themselves the Independent Group, or "IG" for short and snappy.

The word "pop" as applied to art was first used in an essay by Lawrence Alloway. But its first visual appearance in a work of art occurred in a seminal pop painting by Richard Hamilton, *Just What Is It That Makes Today's Homes So Different, So Appealing?* in which the word "pop" appears on a phallic lollipop grasped by a grossly overbuilt muscleman, who seems to be nothing but a cardboard cutout, looming over a small seminude woman holding a Hoover vacuum cleaner, pert nipples covered with pasties, sporting a plastic lampshade on her head.

The evolving "aesthetic of expendability," best exemplified by Les Levine's *Star Machine* in New York, was soon being mass-merchandised to the home furniture market in the form of the "Blow" chair, an inflatable PVC armchair pioneered by the Italian design firm Zanotta, one of a group of small furniture firms to have risen out of the wreckage of postwar Milan. British versions soon followed in rapid succession, including Arthur Quarmby's playful "Pumpadine" armchair, which came in transparent and bright cherry red PVC and could be easily inflated with a cardboard balloon pump—included.

Inflatability combined two key pop elements: flexibility and softness. Both were as influenced by the bosomy sensuousness of Gina Lollobrigida and Sophia Loren (and the glossy, pneumatic pages of *Playboy*) as the eye-popping spreads found in the Italian design book *Domus*. Pop in England and the United States was known as "Anti-Design" in Italy, a movement led by such aesthetic visionaries as Ettore Sottsass, who in 1966 exhibited a line of prototype plastic wardrobes covered in colorful sheets of plastic laminate gaily imprinted with pop and op art motifs.

Obsessed with the mechanics of mass production of imagery, pop artists like Roy Lichtenstein were profoundly influenced by comic books. On the Continent, that action-comic-book fixation was mixed with erotica in the popular sci-fi soft-porn strip *Barbarella*. Directed as a futuristic, cinematic sex romp by French filmmaker Roger Vadim (Brigitte Bardot's former flame) *Barbarella* starred Vadim's current girlfriend, Jane Fonda, idly capering through a dystopian landscape of broken-down transparent bubbles.

Star Trek

January 1965, Paris: Though no more than fifty buyers and journalists could possibly be squeezed into the long, narrow white showroom, sparsely equipped with clean rows of white vinyl cube chairs (one fashion journalist later likened it to a "space station"), the tiny atelier above avenue Kleber throbbed with anticipation. The whole fashion world had been prepared for something momentous—some would later reverently refer to it as a "reformation"—to emerge from the lair of André Courrèges, who after spending twelve years as "star tailor" to the great haute-couturier Cristobal Balenciaga, had finally struck out on his own.

The thirty-eight-year-old Courrèges had launched his own ship with his master's blessing, but that didn't mean that he had any intention of following slavishly in his master's footsteps. Courregès was his own man, of his own generation, and from the moment a dark-skinned, lithe Guadalupan dancer named Gerald Felix bolted out of one of the white vinyl cubical chairs arrayed on the floor, promptly peeled down to his white jersey boxer shorts, and began rhythmically undulating to the soft *obbligato* of African tom-toms (casting a sensuous shadow on a white plastic screen resembling a radar dish), it became palpably clear that this was just what the fashion world needed: a swift kick in the Spandex *derrière*.

"After such a sensational start," reported Gloria Emerson of the *New York Times*, "the wonder was that Courrèges' clothes didn't seem a bore." But Courrèges's first "Space Age" collection, though deliberately drenched with sex appeal, radiated an almost Athenian classical unity. As one awestruck fashion editor later exclaimed, "His dresses seemed *sculpted*, not sewn."

"This," *Vogue* decreed, "is the switch that flicks on fashion."

Three months after his stunning debut (March 1965) for spring Courrèges combined his signature blinding "moonstruck white" raiment with a miniskirt—a fashion first—although the young British designer Mary Quant introduced her own mini at about the same time. Soon to be as established a fixture on the sixties scene were the white slit-tennis-ball goggles sported by Courrèges's slinky, slithering models, whose conspicuous lack of hips prompted the *Village Voice*'s Blair Sabol to accuse "The Lord of the Space Ladies" of harboring a "little girl fixation" that compelled him to force "grown women to dress like little girls."

Courrèges's more profound point was to forge a new ethos of syntho-sex, achieving a sensuous hybrid of Barbarella and Bond Girl. Though Geoffrey Beene pronounced Courrèges a "fraud" and a "sadist," others were seduced by the classical lines apparent in Courrèges's exquisite tailoring, which in *Vogue*'s view evoked heroic images of "a modern girl who has journeyed through ancient Sparta."

Not only did Courrèges's fondness for an androgynous future lead him to create the white kidskin "go-go boot" in which Nancy Sinatra promised to "walk all over you," he conceived of another enduring feminist statement: the pantsuit. Courrèges was the author, for better or worse, of unisex clothing.

Androgyne Andy Warhol was impressed by Courrèges's success. "The clothes are so beautiful," he silkily murmured. "Everyone should look the same. Dressed in silver. Silver doesn't look like anything. It merges into everything."

Pierre Cardin hastily jumped on the Mercury retro-rocket by introducing his own "Space Age" look featuring bullet-shaped helmets and goggles. Emilio Pucci soon followed suit with a new flight attendants' uniform for Braniff Airlines, which forced stewardesses to turn up at the airport wearing clear plastic bubble helmets on their heads—meant to protect beehive hair-dos from the elements. Though the general effect was one of carrying the hair dryer out of the salon onto the tarmac, Braniff had successfully linked itself to the future of aviation: space travel. Served by stewardesses straight out of 2001, every junior executive could feel like Gus Grissom in orbit.

Paris, February 1966: An obscure Spanish architect named Paco Rabanne, whose main claim to fashion fame was his princely status as the eldest son of Balenciaga's first seamstress, made the quantum leap from employing synthetic materials to using plastic itself—hard plastic—as a fashion medium. The thirty-year-old Rabanne showed a collection of dresses, vests, and enormous bib necklaces fabricated—or constructed—of phosphorescent Rhodoid plastic disks strung with fine wire. He and his "architectural" fashions were promptly hailed as the worthy successor to Courrèges, while his shiny plastic sun visors, helmets, and chunky plastic jewelry earned him the title "King Plastic."

Rabanne took a structural approach to fashionable deconstruction. He deliberately set out to "outmode" the seam with a system of wire links that held his quarter-sized plastic rhinestones firmly in place like modern chain mail. Rabanne achieved his unique "castmolded" mod medieval look by using wire and pliers where seamstresses like his mother had toiled for countless hours. Though a Psych 101 explanation was that he yearned to liberate his mother from mindless toil, a less generous interpretation was that his hidden agenda was to throw his sainted mother—and thousands of women like her—out of work.

"The only frontier left in fashion is the finding of new materials," Rabanne proclaimed. "To new materials one must apply new ideas. I had to go back to archaic methods to escape old outmoded techniques." By the end of October (1966), Rabanne was eating up thirty-thousand square-meter sheets of Rhodoid plastic a month, and after opening his own boutique on rue Bergère—which he promised would be "very dark inside with fluorescent furniture; the only light will come from the furniture"—he began boldly experimenting with one-piece molded plastic garments produced in tandem with a line of premolded pop plastic furniture. 'Please don't call me a couturier," he protested. "I am an artisan." And in an elegiac homage to Courrèges, he condescendingly pronounced his brilliant predecessor "a flashbulb. He gave out a brilliant white light that very quickly faded." As would Rabanne, leaving little but the stench of cheesy perfume in his wake.

September 1965: A new mod-fashion boutique called Paraphernalia opened on New York's Madison Avenue between Sixty-sixth and Sixty-seventh Streets. Founded as a "laboratory and showcase for new talent" by fashion entrepreneur Paul Young, whose *obiter dictum* was "fashion should be fun, and above all, not taken seriously," Paraphernalia resembled a modern art gallery more than a conventional retail space. Designed by sleek-simple architect Ulrich

Franzen in bleached white, silver, and neutral woods, the store came equipped with a series of small stages on which models "frugged" during peak hours, disco style.

When the discotheque Electric Circus opened in June 1967, "shaggy hippies unconcernedly perform pagan tribal dances," *Life* luridly gaped, "as magnified images of children in a park, a giant armadillo or Lyndon Johnson disport themselves on the white plastic sculptured expanse of the tent-like ceiling."

The relentlessly trendy Paraphernalia rapidly morphed into a retail Electric Circus, promoting the latest in disco trends with a black vinyl minidress by former electrical engineer Diana Dew, featuring eighteen clear plastic windows that lit up when the hip-mounted "power source potentiometer shrunk to the size of a cigarette pack" was activated by the wearer. The potentiometer, synchronizable to the disco soundtrack, let the dress blink on and off to the beat of a different drummer.

The rising young star in Paul Young's stable was Betsey Johnson, a former editorial assistant at *Mademoiselle* who after earning her art degree from Syracuse University began knitting and selling homemade sweaters to her officemates. She knew she liked making clothes better than drawing them, but without any credits in design she could never get past "the receptionist's secretary at the fortress of Seventh Avenue."

Not until *Mademoiselle* senior editors Noni Moore and Sande Horvits introduced her to Paul Young, who was bowled over by Johnson's fondness for a rich array of industrial-strength materials and fabrics that had never before been considered fashionworthy, did Betsey Johnson get a break. Paraphernalia's brave new world embraced "fabrics you'd spray with Windex rather than dry-clean," Johnson later gleefully recalled. "We were into plastic flash synthetics that *looked* like synthetics. It was: 'Hey, your dress looks like my shower curtain!'"

Johnson's keen interest in innovative, high-tech materials and fabrics dated back to her high school years, when she ran her own dance school and often took the train into New York from suburban Connecticut in search of special-order stretch fabrics for her students' recital costumes.

After joining *Mademoiselle* as a guest college editor in 1964, she had the good luck to be assigned to assistant fabric editor J. D.

White, who happened to be eight months pregnant. Asked to review White's bulging files, Johnson was dazzled by the rich store of information and exotic samples "of what the industrial people used, stuff they were making car interiors out of, fabrics they were lining caskets with, the materials they were insulating spaceships with."

One of Johnson's most successful designs for Paraphernalia was a translucent cellophane evening dress, intended to be worn over a flesh-colored body stocking. Riding the wave of fashionable seminudity that also popularized Hollywood designer Rudi Gernreich's topless swimsuit, Johnson launched a vinyl dress that came complete with a kit of adhesive foil scallops, to be glued on in strategic spots to maintain a modicum of modesty. For her next installment, she produced a plastic slip dress equipped with attach-it-yourself sequin appendages (a bit of a put-on) to be worn with clear vinyl shoes by Herbert Levine. And a see-through blue vinyl strapless minidress accessorized with insert-yourself nylon bust pads (strictly optional).

From Courrèges's white Spartan shifts to Cardin's rocket-girl goggles and Rabanne's plastic chain mail, sixties syntho-visionaries employed unnatural materials to evoke an ethereal, androgynous future of moon missions and space stations. In stark counterpoint to the rough-hewn denim-and-velvet Carnaby Street sixties, or the mud-splattered overalls of Woodstock Nation, pop future fashion looked to the test tube and factory, not the farm or harem, for inspiration. By wrapping thin Twiggy bodies in cellophane and automotive vinyl, Betsey Johnson reveled in a kicky, whimsical Un-Nature that made a point of its sharp break with tradition. The hypersynthetic look was shiny, metallic, sleek, and sexy and was soon to resurface as the glitter look espoused by David Bowie in "Ziggy Stardust."

Back to Nature

With a circumference of just over half a mile, Houston's new Astrodome, arching over a clear 642-foot span, stacked up at four-and-a-half times the diameter of the Pantheon in Rome—the largest dome erected until the twentieth century. Since a grass-covered field was considered a must, a roof that let sunlight in but also cut down on visual glare was very much on the design agenda. The result: a louvered roof, covering nine acres of playing field, composed of thousands of Lucite (MMA, or methyl methacrylate) sheets, to provide optimal light diffusion.

To grow a lawn indoors—shades of Betsey Johnson—a team of botanists recruited from Texas A & M recommended a strain known as Tiffway Bermuda, which could survive at a light level 20 percent lower than nature bestowed. But on opening day, the high-traffic infield was covered in a strange new substance called AstroTurf in honor of the Houston Astrodome. Developed by Chemstrand Corporation of nylon monofilament woven into a polyester backing, with a second backing of vinyl chloride layered beneath to provide extra cushioning, the AstroTurf infield looked so much better (translation: *greener*) than the Tiffway Bermuda transplanted-sod outfield that the real grass was *painted green* before the first pitch.

In an environment offering synthetic weather—6,600 tons of windless air pumped into the void beneath the Lucite "sky" a minute—how could anyone object to artificial grass? AstroTurf possessed a certain cool cachet in southern Texas, where phony-baloney rusticity had become as much a way of life as the vinyl saddle on a mechanical bucking bronco.

"Welcome to the Great Phony Outdoors," *Newsweek* sardonically invited contemporary America, where a thriving market had grown up for synthetic stones produced by Loma Stone Inc. of Los Angeles, promoted as "boulders a 98-pound weakling can plant in his rock garden." After which backbreaking labors, the weakling might even have enough energy left over to play a round at the nearby GOLFOMAT Country Club, an indoor, air-conditioned "country-club environment" where you drove your ball at a computer-controlled projection screen, which digitally signaled how many more yards of driving were needed to reach that virtual green.

Packaged Goods

Packaging—meaning to contain an object itself in a most realistic way—exposes its commonness in a beautiful and relaxed manner.

—Christo

No other artist so fully illuminates the twentieth century's preoccupation with packaging.

—David Bourdon, *Christo*

On opening day (November 7, 1969) some 2,500 visitors paid a modest twenty-cent admission for the privilege of trooping and tripping over a million square feet of what *Time* called "clingy, opaque, icky, sticky polypropylene plastic," a shiny, billowing sheet covering a mile-long stretch of craggy coastline embracing the inlet of Little Bay, some nine miles south of Sydney, Australia.

The twelve-foot-wide sheets of woven plastic fabric, stitched together with thirty-five miles of rope, secured by steel staples into the eighty-four-foot-high cliffs, made the jagged rocks, gaping holes, and sharp crevices beneath appear deceptively soft and elastic. Beneath the wind-billowing swags and folds of fabric, danger lurked—the real risks disguised by the blinding reflectiveness of the plastic surface, so smooth underfoot, so devilishly hard to traverse. The going was particularly treacherous near the tide line, where the plastic wrap was constantly soaked by the surf.

Under the ministrations of the peripatetic Bulgarian-born artist who called himself Christo, the plastic-wrapped mile of beach had all of a sudden become not a natural feature of the landscape but a man-made "thing" of extraordinary beauty. The polyethylene sheeting softly and sensuously hugged the contours of the sand and rocks, smoothing out the jagged edges and imparting to the rough, elemental surfaces the subdued features of an anatomy clothed.

Born Christo Javacheff in Gabrovo, an industrial city in the Balkan Mountains of Bulgaria, up until Little Bay he had mainly confined himself to wrapping buildings and other man-made objects—from the Kunsthalle in the Swiss capital of Bern to, more recently, the Museum of Contemporary Art in Chicago. Yet as a young student at the Academy of Fine Art in Sofia, he had been an active if reluctant member of an agitprop youth brigade, primarily charged with painting huge billboard-sized quotations from Stalin and Lenin on the sides of cliffs looming high above the mountain-bound town.

By journeying halfway around the world to wrap a mile of seacoast in Australia, Christo was in a sense returning to his roots, which he had abandoned not long after the 1956 Hungarian uprising (and its subsequent repression) by fleeing westward, first to Vienna, then to Paris, finally to the United States. Christo's fondness for wrapping things might have been born in those dark days of flight as an enigmatic visual reference to what he had left behind:

a looming "Iron Curtain" that had exerted such a profound effect on his life as a refugee.

After wrapping chairs in plain cloth (the "bones" protruding through the tied swags of fabric in ways oddly evocative of the semidraped Nike of Samothrace), young Christo gleefully graduated to wrapping caches of oil drums along the decaying waterfront of Cologne. From there he turned to wrapping thin air with his monumental 1966 *Air Package,* a colossal column of plastic-impregnated canvas and polyethylene rope, which he vainly attempted to erect in a field outside Kassel, in West Germany, before the $5,000 polyethylene skin (manufactured by a firm in Sioux Falls, South Dakota) split wide open in a ten-foot rent after three hours of inflation, and collapsed in a humiliating heap on the ground.

Wrapping the Sydney Cliffs above Little Bay was a different experience: As a work of art, it subtly played off evolving ideas about nature and artifice, an ancient case of point-counterpoint that the deliberate use of a shiny synthetic material radically enhanced. Sydney art critic James Gleeson surmised that part of this outsized work of art's visceral appeal was grounded in "our uncertainty as to whether we are responding to the beauty of nature or the beauty of art—an opposition that merely adds piquancy to the experience."

Work of art or avant-garde hoax, it was a quite a hands-on experience for the sixty volunteers, mainly students, who spent months rappelling and clambering over the treacherous sea-swept rocks and nearly vertical cliffs knotting and securing rope, sewing fabric, firing staples with ramset guns directly into the looming rock face. Halfway through the installation, gale-force winds ripped up nearly half of the project, forcing the team back to the drawing board for repairs and alterations—tailoring a landscape as willfully artificial as the infield of the Houston Astrodome, hailed by one critic as "the most fully artificial landscape experience this side of the Sea of Tranquillity."

In the fall of 1968, Christo had met John Kaldor, a Hungarian immigrant working in Sydney for a major textile manufacturer. Seized with enthusiasm over the idea of "wrapping" something as seemingly unwrappable as a beach, Kaldor approached Prince Henry Hospital, titular owner of Little Bay, about letting Christo wrap their seascape in plastic. Much to Kaldor's surprise, the hospital's board approved the project, prompting the newly environmentally conscious editorial board of the *Sydney Sun* to warn: "This threatens to result in the biggest blow-around of litter the city of Sydney has ever seen!"

The rival *Sydney Sun Herald,* not to be undone on the environmental front, ominously claimed in a front-page report that "Wrapping Little Bay Could Harm Wildlife," leading to the "death of insects and the killing or stunning of wildlife." But Christo's sponsor, Kaldor, leapt to the artist's defense, dashing off furious letters to the editors of Sydney's major newspapers explaining that Christo had deliberately selected "a polypropylene fabric of a type used to prevent soil erosion, woven loosely enough to permit air and moisture to freely circulate."

The staunchly philistine *Daily Telegraph* wouldn't buy it. "Young fairy penguins caught in rock crevices," the paper prophesied, "will slowly starve to death because their parents have been too frightened by the great mass of plastic."

The Little Bay coast remained wrapped for ten weeks. When the show was over, all imported materials—one million square feet of erosion control mesh, thirty-five miles of polypropylene cable, twenty-five thousand steel staples—were removed, and the site scrupulously restored to its original condition. As to the ephemeral nature of his work—in which the dynamic process of wrapping and unwrapping was as critical as the static image of the package itself—Christo told *Time* with a shrug: "It's not a very permanent world anyway."

Plast-o-Pun

Pop artists took the plasticized landscape vacuformed in the fifties and recycled it in an ironic light. By cannily co-opting synthetic materials despised by the highbrows, pop succeeded in tweaking the bourgeoisie on two fronts: the great mass of people who consumed plastic without thinking about it—and threw it away without thinking about it—and the cultural elite who abhorred its inexorable advance on philosophical, moral, and aesthetic grounds.

To Les Levine, plastic's singular beauty lay in the fact that it made its own comment on the prevailing concerns of modern society simply by virtue of being disposable. That this was the ultimate horror of plastic—that once discarded, it never disappeared—was in all part of the same morbid eco-humor. Pop artists like Andy

Warhol used plastic's sly superficiality to transmit the subversive message that in the airbrushed, plasticized sheen of a Campbell's soup can—mechanically replicated ad infinitum—we find reflected an image of modern, alienated humanity.

Christo, by contrast, elevated social commentary into the realm of the synthetic sublime. In a twist on a twist, he repeatedly demonstrated that plastic packaging can be—if handled with finesse and feeling—an enhancement of the pure state of nature.

10

The Seat of the Plague

Courtesy of Fred Kligman

> I sometimes think that there is a malign force loose in the universe that is the social equivalent of cancer, and it's plastic. It infiltrates everything. It's metastasis. It gets into every single pore of productive life. . . .
>
> —NORMAN MAILER
> *HARVARD MAGAZINE,* 1983

Prophecy

On his way west to cover the 1964 Republican convention in San Francisco, Norman Mailer fell into casual conversation on the plane with an Australian journalist named Moffit.

"Why is it," Moffit demanded, "that all the new stuff you build here, including the interior furnishing of this airplane, looks like a children's nursery?"

Moffit's comment struck Mailer as—perhaps unintentionally—profound. After sadly conceding that indeed "the inside of our airplane was like a children's plastic nursery, a dayroom in a children's ward," Mailer was uncharacteristically struck dumb. But, as he would later recount in *Cannibals and Christians*, had he been Quentin Compson (Faulkner's hero in *The Sound and the Fury*), he might well have shouted out:

> Because we want to go back! Because we developed habits which are suffocating us to death. I tell you, man, we do it because we're sick, we're a sick nation, we're sick to the edge of vomit and so we build our lives with materials which smell like vomit, polyethylene and bakelite and fiberglass and styrene. . . .

The interior of a jet airplane, the author ominously mused, "is after all one of the extermination chambers of the century. Slowly the breath gives up some microcosmic portion of itself. Green plastic and silver grey plastic . . . nostrils breathe no odor of materials which existed as elements of nature, no wood, no stone."

Under closer inspection, the "children's plastic nursery" had monstrously evolved into something far more lethal and sinister: the late-twentieth-century capitalist equivalent of the gas chambers at Auschwitz.

Throughout his brief sojourn in San Francisco, Mailer found himself dogged every step of the way by the specter of the demon plastic. In his room at a spanking-new Hilton:

> The carpets and wallpapers, the drapes and the table top were plastic. The bathroom had the odor of burning pesticide. It developed that the plastic cement used to finish the tiling gave off this odor during the month it took to dry. Molecules were being tortured everywhere.

In Mailer's grim equation of modern life, plastic equaled vomit, contraception, cancer. Its routine ascendancy could be regarded, metaphorically at least, as responsible for the degraded state of public discourse, for air, water, and land pollution, for a collective specieswide regression to an infantile state, and—come to think of it—the rank odor of inauthenticity pervading modern society.

From "NUTS," an editorial published in *Modern Plastic* (March 1971):

> Norman Mailer, renowned author and unsuccessful candi-date for mayor of New York, recently offered a new gospel on plastic. The occasion: an interview with Arlene Francis on New York radio station WOR.

> What is responsible for "this revolutionary generation"?
> The plastic environment.
> What causes drug addiction?
> Plastic nipples.
> What pollutes rivers and lakes? You guessed it.
> What is the "infinitely painful expression of the devil?"
> Right again.
> What kills "the soft surface of youthful skin?" Exactly.

> The "blame it on plastic" game is one that people have been playing for a long time, and the plastic industry has learned to live with it. The players usually come from the

nether regions of vested interest, ignorance, or both. But it's another thing when a person of Mailer's stature gets access to the airwaves to broadcast such patently irresponsible nonsense.

How about equal time? How about getting the plastic story on the air, by telling of the many great benefits that plastics have brought to mankind? In the meantime, let's take comfort in the final thought Mailer expressed at the end of the interview: "I hate to sound like a nut."

You said it Norm, not us.

Proof

Norman Mailer may have sounded like a nut to the plastics industry, but by the early seventies the much disputed link between cancer and some plastics had been well established by respected epidemiologists. The plastic-cancer equation could no longer be considered a paranoid fantasy spouted by the lunatic environmental fringe. Still, the plastics industry resolutely refused to give credence to the idea that plastic had a few things to answer for. Stonewalling was the preferred strategy.

From *Science News,* September 7, 1974:

TOXIC SURPRISES FROM THE PLASTICS INDUSTRY
. . . 2.5 million workers are engaged in producing 29 billion pounds of plastic each year in the U.S. But one big yellow streak runs across this rosy picture . . . many of the chemicals used to make plastic are so toxic they affect workers' health.

. . . 15 vinyl chloride workers recently died from a rare chemically induced liver cancer . . . while OSHA (Occupational Safety and Health Administration) toxicology studies of compounds used in plastic manufacture reveal that . . . exposure to small but still undefined quantities of vinyl chloride can cause fibrotic lesions on the liver after so much as a year of exposure. . . .

From *Science News,* March 1, 1975:

TOXIC FUMES FROM FIRE RETARDED FOAM
At the Flammability Research Center at the University of Utah, three scientists (H. J. Voorhees, S. C. Packham, I. N. Einhorn)

exposed rats to fumes of polyurethane foam treated with common phosphate fire retardant. After a twenty-minute exposure to the smoke, all exhibited severely impaired ability to perform normal movements. After an hour, many rats developed *grand mal* seizures . . .

Rats exposed to burning foam not treated with flame retardant were able to escape from a plate-sized circle in six seconds, while rats exposed to burning foam treated with flame retardant became so disoriented and convulsive that they could not move out of the circle in less than a minute.

If Norman Mailer was the 1970s antiplastic movement's leading literary light, ecologist Barry Commoner was its Rachel Carson. In Commoner's blistering 1971 antiplastic polemic, *The Closing Circle,* plastic and related synthetic materials ominously filled the mustache-twiddling ecovillain role once played by the deadly pesticide DDT.

"Synthetic chemicals, specifically chlorinated organic compounds," Commoner contended in an interview conducted deep inside the belly of the beast (*Modern Plastics'* midtown-Manhattan offices), "are responsible for the appearance of dioxins in the environment." To those unfamiliar with the term, Commoner defined dioxins as "vicious carcinogens linked to a marked increase in the use by industry of chlorinated chemicals—specifically PVC [polyvinyl chloride]." PVC is more popularly known as vinyl.

After plastic's carcinogenesis, its greatest sin, environmentally speaking, was its apparent imperviousness to organic biodegradation. What had once been regarded as its unique virtue—extraordinary durability—had suddenly emerged as its singular vice. Every piece of plastic ever manufactured in the Synthetic Century continued to haunt a vulnerable natural environment. Sea turtles that chewed on floating polyethylene film garbage bags, in the mistaken belief that they were jellyfish, choked by the thousands. In the South Atlantic's wide Sargasso Sea, sea animals routinely ingested raw (unprocessed) plastic pellets, mistaking them for plankton. At one location in the North Pacific, some fifty thousand northern fur seals died every year after becoming entangled in discarded nylon mesh fishing nets. Nets which, after being set adrift in the open sea, neither sank, broke, nor rotted away but continued to drift across the oceans, possibly for centuries, harvesting phantom catches into a demonic eternity.

In Commoner's view, the image that best symbolized the sinister role of the synthetic web in the modern ecocatastrophe was a widely distributed photograph of a wild duck, "its neck garlanded with a plastic beer-can pack," which having been "too trustingly eager of modern technology," he noted grimly in *The Closing Circle*, "had plunged its head into a plastic noose." A noose no doubt destined one day to catch up with the rest of us if we blithely refused to change our plastic-happy ways.

Like a sci-monster in a cheap horror flick, plastic had metamorphosed in the contemporary environmental drama as the great Eco-Satan. To a public only lately aroused to a state of acute plastiphobia, the confirmation of links between cancer and plastic only accentuated more generalized fears about the dubious role of other synthetic compounds in a wide variety of human ailments.

To a beleaguered plastics industry, incineration held out the greatest hope of recovering some of the energy so extravagantly expended in manufacturing the latest generation of petroplastics. But even if equipped with the best air-pollution-control equipment available, few ecologists believed that combustion—even at extremely high temperatures—could ever be made truly safe. The plastic industry's last great hope ultimately loomed as ecologists' worst nightmare: Combustion, as opposed to dispensing with the dread carcinogens, would by converting them into more insidious forms—gases or particulate matter—disperse them more widely across a syntho-saturated planet.

That burning plastic could release noxious gases equal to any biochemical weapon devised at Fritz Haber's Kaiser Wilhelm Institute in Berlin was dismally proven in January 1970 when an elderly resident of the Harmar House nursing home in Marietta, Ohio, carelessly tossed a lighted cigarette into a polypropylene wastebasket filled with waste paper.

The burning paper caused her plastic basket to flare up, throwing out flames that rapidly consumed her polyurethane foam mattress, touched off her nylon wall-to-wall carpet, and instantly ignited the carpet's styrene-butadiene foam underlayer. By the time rescue workers arrived on the scene to evacuate the ward, they were met by a dense, black wall of smoke that obscured their view of survivors still trapped inside. The billowing smoke not only blinded the firemen but was so viciously toxic that it overcame scores of enfeebled patients who might otherwise have been able to escape on their own. By the time the fire was brought under control five hours later, twenty-two elderly people had died. The vast majority, coroners concluded, had been felled by the toxic fumes, not the flames.

Eight months later, on a steamy day in August 1970, a twelve-alarm fire broke out on the thirty-third floor of One New York Plaza, a modern high-rise office building in Manhattan's financial district. A stray electrical spark ignited a welter of computer cables concealed within a dropped ceiling in a telephone equipment room, which was itself filled floor to ceiling with mile after mile of exposed polyethylene-insulated cable. As the heat intensified, flammable and toxic gases were distilled from the polyurethane foam padding cushioning the office furniture in the suites below. As the toxic gases burned, the blaze exploded, as if shot out from an aerosol can. Fed on this rich diet of toxic, flammable gas, the fire consumed two entire floors covering over forty thousand square feet of office space in under twenty minutes. During the six hours it took to extinguish the flames, two firemen died of smoke inhalation. Thirty more were hospitalized with potentially life-threatening lung injuries—as the result of inhaling burning, noxious plastic fumes.

Only three weeks later, at 8:30 on the morning of August 26, a third blaze broke out in the recently completed British Overseas Airways terminal at New York's John F. Kennedy Airport. As the flames licked across six hundred polyurethane foam-padded benches clustered by the gate entrances, clouds of toxic gas distilled from the benches' foam padding caused the fire to gallop off down the 35-foot-wide, 330-foot-long corridor at lightning speed, "like a giant gas jet burning fiercely at ceiling level," one airline safety official later recalled. As the roaring fire leapt wildly from seat to seat, blowing out dozens of large plate-glass windows in its wake, it took a mere fifteen minutes to consume the entire west gallery of the newly competed airline terminal, at an estimated cost of $2.5 million in damages.

Awestruck insurers would later term it "the shortest large-loss fire in the history of mankind." Rexford Wilson of Firepro Inc., a flammability expert retained by BOAC's insurance company to investigate the blaze, delivered the bad news to plastic manufacturers on the burning issue of plastic fire safety:

"Plastics ignite like excelsior, contribute heat like kerosene, and produce four to thirty times the amount of smoke as nonsynthetic materials."

Occupational Hazard

When polyvinyl chloride (PVC) was first introduced in the United States by BFGoodrich in 1929 (developed by Waldo Lonsbury Semon after watching his wife cut a shower curtain from non-water-resistant cotton cloth), it was widely hailed as a modern miracle. Cheap, chemically stable, fire resistant, and able, as *Fortune* put it, "to assume an extraordinary range of soft and hard forms," vinyl proved an unalloyed blessing for car manufacturers, record producers, and designers of vinyl topped "convertible" cars.

Unfortunately, the same could not be said for the workers who handled it. For decades, the only significant occupational hazard associated with handling vinyl chloride gas—a key component in the manufacture of vinyl—was a tendency for workers to occasionally get high off the fumes. In fact, PVC gas's narcotic properties were so overwhelming that public-health authorities briefly authorized its use as an all-purpose anesthetic until it was found to cause heart arrhythmia. But halfway through the vinyl-happy fifties, reports began filtering in from as far afield as the Soviet Union that high levels of VCM (vinyl chloride monomer) exposure might predispose workers to a rare hepatitis-like condition. Equally alarming, thousands of European vinyl workers were found to be suffering from a wide range of maladies including gastritis, skin lesions, and dermatitis. Occupational safety authorities began to speak tentatively of a new "iatrogenic" (man-made) disease, which they called "vinyl chloride syndrome."

Gastritis, dermatitis, and even hepatitis were all unpleasant disorders, but not typically life-threatening. But in 1961, an experiment conducted by Dow Chemical revealed that rats exposed to VCM levels as low as a hundred parts per million had developed liver defects and tumors. Vinyl had at last been linked, although not yet conclusively, to the Big C.

In response to these disquieting findings, Dow took the unusual step of voluntarily tightening the permitted VCM-exposure levels at all of its vinyl plants to levels averaging as low as a hundred parts per million—approximately five times lower than the level certified as safe by occupational safety authorities.

But Dow did not widely disseminate its evidence of a vinyl-cancer causation pattern, preferring to do what little it could to protect its own workers without raising a national cancer scare.

No further progress was made in establishing beyond a reasonable doubt the emerging vinyl-cancer connection until 1970, when Dr. Pierluigi Viola, an Italian researcher and expert in acroosteolysis (a rare disease of the finger bones often detected in vinyl workers), announced that rats exposed to VCM levels of thirty thousand parts per million for up to a year had developed angiosarcoma of the liver—a rare cancer affecting the liver's blood cells. Viola's small-scale results were quickly confirmed by Dr. Cesare Maltoni, an internationally respected Italian researcher, who had been commissioned by the chemical conglomerate Montedison to conduct large-scale animal experiments on vinyl toxicity. By 1973, Dr. Maltoni announced that enough rats exposed to vinyl gas had developed angiosarcomas of the liver at prolonged VCM exposure levels as low as 250 parts per million. The results of this study were conclusive enough to put the world's vinyl makers on notice that prolonged exposure to VCM gas was likely to cause cancer in workers exposed to it over extended periods.

Within months of Maltoni's announcement, Dr. John L. Creech, a surgeon employed part-time as plant physician at BFGoodrich's vinyl plant in Louisville, Kentucky, learned that one of his workers had recently died of angiosarcoma of the liver. Recalling that a second worker employed at the same plant had died of the same disease a year before, Creech notified his superiors at Goodrich. Before they could even launch a full-scale internal investigation, a third vinyl worker died from the same condition. To Creech's horror, Goodrich soon learned that a total of eight workers employed at the Louisville plant had been diagnosed with liver cancer, all facing near certain death. Goodrich had little choice but to sound the alarm. The company promptly notified NIOSH (the National Institute of Occupational Safety and Health) of the disturbing pattern of events. NIOSH quickly dispatched a small army of inspectors down to Louisville.

Rather than wait for their own tests to be concluded, NIOSH immediately issued a "preliminary," "emergency," "temporary" VCM-exposure standard, requiring manufacturers to maintain VCM levels at under fifty parts per million. By setting this standard somewhat arbitrarily, NIOSH left itself open to court challenge, and set the stage for a pitched battle between industry advocates, who pledged—with a great show of reluctance—to abide by the new standard on a permanent basis. Occupational safety experts, health professionals, ecologists and labor representatives, however, were firmly convinced that even fifty parts per million was too high.

The phrase "zero tolerance," so blithely bandied about by public health and occupation safety experts, became a negative rallying cry raised by an increasingly defensive plastics industry in protest against the imposition of technically unsustainable environmental controls. "Even a system that doesn't leak, leaks!" thundered Firestone president Todd Walker at the OSHA hearings, before proceeding to elaborate in excruciating detail each and every capital expenditure that would be required—or so he claimed—to reach such an utterly unrealistic goal. If forced to abide by this absurdly risk-averse standard, Walker openly threatened, Firestone would be forced to pull out of the vinyl business altogether.

To such blatant blackmail AFL-CIO health director Sheldon Samuels righteously replied: "The men and women we represent will not countenance the barbaric attitude which seems to dictate that death and disease are all just part of the sacrifice that must be made for food, clothing and shelter!"

Still, even the Feds were loath to pronounce an economic death sentence on such a politically powerful industry if they could figure out some way around it. But given their clear mandate to protect worker safety, it was hardly surprising when OSHA and NIOSH sided with the occupational safety authorities by issuing a ruling mandating "no detectable level" or "zero tolerance" for carcinogens in the workplace. There appeared no compromise that would satisfy both parties.

While the Society for the Plastics Industry took NIOSH and OSHA to court, after months spent protesting the costly protective measures being pushed by NIOSH—which mainly consisted of providing vinyl workers with airtight suits and respirators—within days of the "zero tolerance" ruling, a cluster of industry leaders (including Dow Chemical, Union Carbide, and Goodrich) took steps to reassure anxious stockholders that they had no intention of cutting back on vinyl production. Providing, of course, that they could get their hands on enough respirators to keep the vinyl vats flowing. The Feds had called the industry's bluff, and thousands of life-threatened workers were better off as a result.

In the wake of the pitched vinyl battle, the Federal Bureau of Alcohol, Tobacco and Firearms (BATF)—which had been widely expected to approve a PVC liquor bottle after three years of probational marketing—executed an abrupt about-face, pulling the plug on all plastic food packaging not yet approved until it could be proven safe for products slated for human consumption. What enraged PVC-bottle manufacturers about the BATF ruling was that they had—or so they claimed—made substantial progress in eliminating even the most minute traces of vinyl monomer in PVC food packaging. In fact, no test was sensitive enough to detect any carcinogenic vinyl monomer leaching from PVC bottles into contained liquids—this was the dread "monomer migration." But an environmentally aroused Congress, preferring to remain safe rather than sorry, had recently passed the "Delaney Clause" (as an amendment to the Federal Food and Drug Control Act) expressly forbidding the presence of any known or suspected carcinogens in all food products and packaging sold within the United States.

The Bottle Wars

No lily-livered cancer scare was going to keep a proud plastics industry from achieving its decades-old dream of distributing soft drinks and other carbonated beverages in nonreturnable plastic bottles. By the mid-seventies, the venerable system of distributing soft drinks in single-serve, glass, "two-way" refillable bottles had broken down and was not about to be replaced.

The new beverage industry buzzword was "convenience." Convenience dictated that soft-drink bottles grow from king to family size, from single to multiserve, and that the new generation of "one-way" nonrefillable bottles be willy-nilly disposable. This meant chucking the empty wherever and whenever the consumer pleased, without regard to ultimate destination, whether a crowded landfill, or gleaming on some once verdant roadside, or gradually—very,

very gradually—deteriorating on some bottle-strewn inner-city mean street.

Bottlers couldn't have been happier with the new, hefty sixty-four-ounce nonreturnable, resealable glass bottle, even if it was not so convenient or commercially viable for municipal authorities charged with garbage disposal and litter collection. The kingsize bottle was a huge hit with consumers because it was easy to pop in the back of a station wagon and held enough soda to keep a clutch of kids sugar-satiated for hours. As one savvy soft-drink distributor gleefully gloated, "Put a two-liter bottle in a consumer's hand and she'll stick it in the fridge and serve it all day." It might not have been good for the public or national health, but it was great for the wealth of the bottlers and beverage makers.

The new family-size bottle had only one problem: It was built like a bomb. It had to be, because the bigger the package, the thicker the glass walls needed to be to withstand the enormous pressure—up to four atmospheres—under which carbonated beverages are stored. "That family-size glass package was a lethal weapon," one soft-drink distributor grimly recalled. "And the bigger and heavier those glass bottles grew, the more worried we got about consumers being injured by flying shards of glass." Glass had become a liability—a product liability.

The first solution to the weight and breakability challenge was the disposable aluminum can. But its sharp-edged flip-top caught in the craws of birds and other wildlife and lacerated their throats. Besides, consumers preferred bottles—multiserve bottles—of which the only truly safe kind was plastic. So what if vinyl hadn't made the grade?

There were plenty of other polymers waiting eagerly in the wings, poised to swoop down and turn glass into history. Two major plastic manufacturers, Monsanto and Du Pont, were each preparing to go to market with a contender: one made out of polyester, the other of acrylic plastic. Coke's candidate was acrylonitrile styrene, which Monsanto had developed in conjunction with the Coca-Cola Company at an estimated cost of $100 million. Compared to today's soda bottle, the first plastic Coke bottle was stiff and brittle, and was intended to be a close facsimile of the returnable, refillable, two-way green-tinted glass "contour" bottle (derived from the shape of a coca-bean pod) Coke had come in for nearly a century. Pepsi, meanwhile, had been working with Du Pont on its own polyester-based version of its equally familiar swirl-patterned bottle, made from a new film called PET.

Unfortunately for Coke, buried deep inside Monsanto's petition to the FDA requesting approval for its brand-new acrylonitrile Coke bottle lurked (according to one published report) "an indication that extremely low levels of carcinogenic monomer may form and migrate into the beverage." As it turned out, those "extremely low levels" were on the order of fifty parts per *billion*. But even that wasn't low enough to satisfy a jumpy FDA, still reeling from the PVC scare. When a series of independently conducted animal tests confirmed that a population of rats fed on a steady diet of ANS had developed birth defects and tumors, the FDA chose to conclude its report on acrylonitrile styrene with a classic waffle: "Migration may reasonably be expected under the intended use of a food packaging substance, even though migration of the substance is not analytically detectable."

The saga of the first plastic Coke bottle concluded on a true Watergate-era note, when a scientist performing independent tests on the new Coke bottle—a beverage-industry Deep Throat—tipped off syndicated investigative columnist Jack Anderson that the new bottle might have a cancer problem. This was just hot enough for Jack Anderson to handle: to blow the lid off the Coke-Monsanto plastic-liver-cancer conspiracy. A week after Anderson's report hit the stands (February 11, 1977) the FDA issued a revised directive stating that a new "monomer migration" level of fifty parts per billion had been set for beverage-bearing plastic soda bottles. Coincidentally, the new plastic Coke bottle currently under consideration had failed—by a hair—to meet that tightened standard. A week later, a short obituary for the first plastic Coke bottle was buried on page thirty of the business section of the *New York Times*.

MONSANTO CLOSES PLASTIC BOTTLE UNIT OVER FDA REPORT

The Monsanto Company shut down three plants yesterday that manufacture plastic bottles . . . because of the uncertainties created by a recent Food and Drug Administration statement on plastic bottles. A week ago, the FDA backed down on its controversial approval of the use of acrylonitrile plastic bottles for beer and soft drinks because of studies that showed the material might cause lesions or other growths in test animals.

The FDA never did officially ban the bottle. It didn't have to. The mere prospect of sparking a cancer scare was enough to cause Coke and Monsanto to pull the plug.

Stupid Pet Tricks

In the wake of the plastic Coke bottle debacle, the winner by default of the bottle wars was Coke's arch rival Pepsi's PET (polyethylene terephthalate). The brainchild of Nathaniel "Nat" Wyeth—eldest son of prominent illustrator N. C. Wyeth, older brother of popular artist Andrew Wyeth, and highly regarded polymer engineer at Du Pont—PET was precisely what the plastic industry needed to prosper in the soft-drink market: a safe, strong, unbreakable container that would in time replace glass on all but the most special occasions.

Before his retirement, Nat Wyeth had been known to boast that his job as a polymer engineer was a lot harder than that of his more famous artist brothers and father, because while a visual artist has only to imagine a picture and paint it, a polymer engineer has to imagine a new synthetic substance, develop it, test it, and make it work. And then, in this risk-averse day and age, hope that it passed muster with the regulatory authorities.

One day in the mid-fifties, Wyeth filled an ordinary milk jug made of high-density polyethylene (HDPE) with Coke and left it in his lab refrigerator. After opening his fridge the next morning, he found—not much to his surprise—that it had exploded during the night like a bomb. After cleaning up the sticky, gooey mess smeared over his fridge's insides, the good doctor embarked on a systematic analysis of the one polymer he felt might be able to stand up to the pressure of holding a carbonated beverage: poor, much maligned polyester.

Even when cast in film form, garden-variety polyester was simply not strong enough to keep a bottled carbonated soft drink from behaving like a ballistic missile. But Wyeth knew from experience that stretching a polymer could yield a marked increase in tensile strength. So after taking a length of polyester yarn and stretching it, and finding that its tensile strength dramatically increased, he took a piece of polyester film and stretched it "biaxially"—in two directions, lengthwise and widthwise. He thereby succeeded in "reorienting" the polyester molecule so as to boost its structural strength to an entirely new level.

PET became the first plastic strong enough to hold highly pressurized carbonated beverages without bursting, and safe enough to pass muster with the FDA. By late 1976, capitalizing on Coke's misfortune, PET and Pepsi staged a swift blitzkrieg conquest of the American soft-drink market. Manufacturers embraced PET for its light weight, strength, and unbreakability. But environmentalists were outraged by its advent as yet another sorry example of the plastics industry's insatiable drive for market share, until the nation's dwindling landfills were choked by a tidal wave of plastic discards rampantly spreading like some alien virus across the surface of the planet. As Norman Mailer grimly opined, "They'll go on forever, some of us hope and some of us don't hope. But those that do are capturing the world."

Banning the Bottle

Its cancer problems solved, environmental issues quickly moved to the fore of the PET bottle's uncertain future. The widespread acceptance of the nonreturnable PET soda bottle, by beverage bottlers and the public alike, had the unanticipated effect of supplying environmental activists with just the ammunition they needed to push for a national plastic bottle ban—or, in lieu of an outright ban, a bill regulating its use.

Proposed in various forms over the years—but never passed—a national bottle bill would have levied a straight tax or a returnable deposit on all disposable containers, aluminum and plastic, sold in America. But the beverage and bottling industries put up stiff resistance, coordinated by chief cheerleader Richard Lesher, president of the U.S. Chamber of Commerce.

"You wouldn't ask people to store garbage in their refrigerator, would you?" Lesher sputtered. "But that's what you would be asking them to do—to return soiled containers to the supermarket, store them out back and move them all through the food system. You're going backwards environmentally. You're going backwards in terms of health and hygiene. . . . "

With rhetoric like that, the superbly politically savvy beverage industry succeeded in bottling the bottle bills up in Congress for the next twenty-five years. That legislative vacuum left state and local authorities to weave a motley collection of sanctions, taxes, and levies on the surging synthetic sea. Still, like a Styrofoam cup bobbing eternally somewhere in mid-Pacific, the solid-waste crisis would not go away, no matter how deftly plastic-industry lobbyists spun a tale of woe and discrimination and left-wing conspiracy against their desirable products—a war of words conducted by legions of faceless, bloated bureaucrats.

By 1989, two decades after *Modern Plastics* first editorialized that the issue of waste disposal would be the industry's undoing, even staunchly pro-business *Fortune* was chiding the chemical industry for so steadfastly sweeping its responsibility for the depths of the solid-waste crisis under the rug. The result of such skilled evasive tactics was a welter of antiplastic bills proposed in localities around the country. *Business Week* had one word of advice for our old friend Benjamin, circa 1989: "Stay out of plastics, young man—they're being restricted in Minneapolis—and threatened elsewhere." Not only did Minneapolis's city council ban nonrecyclable plastic food containers by a lopsided 12–0 vote, but the ban sailed through despite a well-organized effort by the plastic industry urging residents to phone council members and tell them "what they felt."

What the callers felt was not exactly what the plastic industry had in mind. Instead of mounting an emotional campaign in defense of their right to eat hot burgers out of plastic clamshells and drink cold soda out of plastic bottles if and when they liked, an overwhelming number of callers strongly favored the ban.

The plastic soda bottle ultimately achieved its greatest potency as a symbol of the increasing depth—and width—of the American environmental divide. To the pro-plastic forces, the commercial triumph of the plastic bottle was a sure sign that the public "loved" plastic, and even more importantly, that it was their free-market privilege to consume whatever container happened to be most convenient. The environment, in other words, be damned.

To environmentalists, the PET bottle was another particularly egregious example of the insidious creep of Barry Commoner's "synthetic web" across a commercial landscape already strewn with the symbols of imminent ecological collapse. To its detractors, the plastic bottle heralded an impending apocalypse; to its supporters, it was yet another sign of chemical progress toward a safer, easier world in which a dropped bottle did not portend a catastrophe in the home.

Yet even to its fiercest opponents, the plastic soda bottle was convenient. It offered a convenient way to demonstrate which side you were on—that of the eco-devils or angels, the villains or heroes, the white or black hats. As even *Audubon* conceded, the often arcane debate between ecologists and industrialists on the relative merits of plastic versus paper garbage bags in time acquired all the sophistic circularity of medieval theologians wrestling over the number of angels that could fit on the head of a pin.

Meanwhile, the reigning high priest of the antiplastic party, Norman Mailer, publicly speculated that the very existence of plastic—in combination with television—might predispose Americans to "get us much closer to totalitarianism than the FBI or the CIA ever could."

As Mailer mused: "If you wanted to convince someone of something that would be very hard for him to swallow, wouldn't it be a good idea to half anesthetize him first? And plastic does that. It just deadens us. ... I think one of the reasons cocaine use is so widespread now is that people's nerve ends are so deadened that they need something to absolutely jack up those nerve ends."

The Devil and the Styrofoam Cup

As a young college graduate in the late sixties, Karl Strehle had been on the verge of accepting an overseas assignment with the Peace Corps when, at the last moment, he opted to stay home and teach. His reasoning at the time was that education would provide a more direct route to achieving his ultimate goal: "making a difference in society."

After spending the next twenty years teaching social studies to junior high school students in West Milford, New Jersey, however, Strehle was ready to take a break. He felt so utterly burned out that in a fleeting moment of weakness he considered "going into business." Fortunately for Strehle and for business, the West Milford New Jersey School District offered him a reasonable alternative. He was given a

Courtesy of Fred Kligman

chance to teach high school, which gave him a new lease on life.

"It was the best thing that ever happened to me," Strehle later told the *New York Times*. "These kids have been such an inspiration to me. Thanks to projects like this—really, this is what teaching is all about. It's getting involved in things in society."

The project that reignited Strehle's flagging enthusiasm for social causes was inspired by a lecture he gave on the American revolutionary Thomas Paine. After listening to Mr. Strehle hold forth on

Paine's capacity to undergo grave personal injury in the cause of social justice, a bright fifteen-year-old sophomore named Tanya Vogt went home and researched a term paper. When she read it aloud in class two weeks later, it was in style and content an echo of Paine.

Rather than eloquently rail against the inequities of colonial government, Tanya Vogt chose the most detestable thing in her own social sphere to protest: the presence, right in her very own school cafeteria, of toxic and ozone-depleting styrene foam cups and trays. After being wastefully used by the barrel, at the end of the day they were just chucked away—to clog up local landfills, forever.

Tanya Vogt had done her homework. She told the class that the chemicals used to blow styrene plastic into a foam were the same ozone-depleting chlorofluorocarbons used as accelerants in aerosol cans, until they were banned for their catastrophic environmental effects. Equally horrifying, Styrofoam cups and platters were the last possible plastic substances to biodegrade even among the beleaguered battery of despised synthetics. The worst part of this styrofoam's saga was that styrene materials, for the most part, were used for a few seconds or minutes—to keep food and liquids insulated—and were then routinely discarded. But in actual fact, they never went anywhere. They just hung around, poisoning and leaching and doing untold damage, wherever and whenever they ended up.

Like Thomas Paine, Tanya Vogt was no pie-in-the-sky social theorist, but a hardheaded activist armed with an agenda designed to foster change on a local level. The basic thrust of Vogt's argument was that to save the ozone layer, the cafeteria of West Milford High School should start using paper instead of plastic trays and cups. Cost was an absurd excuse for openly and blatantly poisoning the environment with toxic chemicals.

Inspired by Tanya Vogt's ardor, Karl Strehle seized the initiative by urging all twenty-seven students in his class to embrace her cause as their own. He even handed the class a plan of action: If faced with such a dilemma, Strehle confidently asserted, Thomas Paine would have organized a lunch boycott to force the authorities to mend their errant ways.

An aroused Tanya Vogt organized a core group of two hundred students, who began exerting subtle peer pressure on fellow students to cough up the extra nickel required to serve lunch on paper plates and trays instead of plastic foam. Some students even gave out nickels to win over key holdouts. Other students were given nickels by students they didn't even know.

Vogt's anti-Styrofoam campaign spread "like a chain reaction" across the state, she later recalled, moved by the depth of plastiphobia her campaign had tapped in the idealistic youth of the affluent New Jersey suburbs. To help get the word out, she organized demonstrations in front of McDonald's restaurants across the state, at which she was repeatedly photographed brandishing placards proclaiming: "Ronald McToxic Uses 100% Non-biodegradable toxic polystyrene which adds to atmospheric pollution . . . 31.6 billion styrofoam clamshells sold!" As a result of this and other successful grass-roots efforts, Vogt was invited to address a Model Youth Forum at the United Nations headquarters in New York.

In a scene that must have sent shivers through the Society of the Plastic Industry, Christopher J. Dagget, New Jersey state commissioner of environmental protection, paid a special visit to West Milford to honor the students for their courage in fighting back the rising tide of plastic foam. He even promised to discontinue drawing from an ozone-depleting stockpile of foam products stored in his environmental department's own cafeteria in Trenton.

Twentieth-Century Disease

With a tornado predicted to hit northern Illinois that afternoon, Theron G. Randolph, a young Harvard-educated allergist and immunologist, was hardly surprised that all his patients canceled— but one. She was a retired cosmetics saleswoman and chronic sufferer from persistent headaches, nausea, dizziness, blackouts and fatigue, who managed somehow to haul herself into Dr. Randolph's experimental allergy clinic in Aurora, Illinois, in the late spring of 1951—the peak of the pollen season.

Randolph's desperate patient had been forced to give up a job she loved because she suffered so much on the road. As punishment for seeking treatment, she had been repeatedly classified as hypochondriacal or neurotic by conventional physicians. But Randolph "believed" in her illness, and with the whole afternoon to devote to her condition alone, persistent questioning elicited from her an intriguing thread: fossil fuel. At her job, and on the road, this woman was routinely exposed to the whole roster of petroleum by-products, including gasoline, fuel oil, auto exhaust, fingernail polish, and a vast array of cosmetic products, many of them derived from coal tar.

The year before, Dr. Randolph, a graduate of the University of Michigan Medical School, former research fellow in allergy and immunology at Harvard Medical School, and faculty member at Northwestern University Medical School, had permanently broken with the immunological and medical establishments by establishing an "environmental isolation" unit in Aurora. The purpose of "environmental isolation" was to systematically detoxify people from the lethal chemical compounds spawned by the Synthetic Century. Randolph firmly believed in a condition he variously called "environmental illness," otherwise known as "twentieth-century disease."

In 1962, the same year that Rachel Carson's *Silent Spring* achieved best-seller status, Randolph published—to far less acclaim—his landmark study, *Human Ecology and Susceptibility to the Chemical Environment,* in which he argued that countless illnesses and allergies suffered by thousands of people are the result of exposure to everyday and often commonplace chemical poisons—the vast majority petroleum or coal-tar derivatives. In other words, synthetic substances.

Leading the list of materials that Randolph was convinced exuded noxious vapors and fumes—or in the parlance of the new clinical ecology, "outgassed"—was plastic. With clinical training to go by, Randolph had come to roughly the same conclusion Norman Mailer had arrived at through intuition: that (to quote Mailer) "there is a malign force loose in the universe that is the social equivalent of cancer, and it's plastic." Randolph firmly believed (and was drummed out of the orthodox medical establishment for holding to that belief) that plastic "infiltrates everything. It's metastasis. It gets into every pore."

A month after being diagnosed by Dr. Randolph as "chemically sensitive," Lynn Larson, a former nurse, awoke from a deep, drugged sleep with the noxious odor of formaldehyde emanating from her pillowcase. The fumes clogged her nostrils, creating one of the worst headaches she had ever experienced, and only confirmed what Dr. Randolph had told her: that formaldehyde, a suspected carcinogen, frequently "outgassed" from synthetic materials like her cotton-poly blend pillowcase. After replacing her pillowcase and her other bed linen with pure, unbleached cotton, the headaches miraculously vanished. As Larson vigilantly concluded in a book she wrote on environmental illness: "One's sense of smell is invaluable in defending against possible chemical injury from modern synthetic products."

Every morning, after reading the newspaper, Lynn Larson's husband would insert it into a zippered nylon mesh bag and bake it in the oven for forty minutes. That way Lynn could touch the chemical ink without being subjected to the dreaded "outgassing." Larson's new lifestyle required turning the clock back to the thirties, to an era before the widespread use of synthetic materials, to a time when, as she said, "we embraced products that I remember my mother using."

Other sufferers from "twentieth-century disease"—others preferred EI/MCS, for "environmental illness/multiple chemical sensitivity—were worse off than Larson. Tony Jones, the Welsh-born former head of the Art Institute of Chicago, had been a successful sculptor working in fiberglass and other synthetic resins and polymers when one day he collapsed in his studio.

After being revived by one of his students—felled by toxic fumes to which he succumbed despite his habitual use of a double-nozzled gas mask—Jones turned his back on his art, and turned in his sculptor's chisel for an activity less hazardous to his physical if not his mental health: art administration.

The obscure British pop singer Sheila Rossall, thirty-one, shriveled to less than fifty-four pounds after confining herself for three years to a single dark room in her apartment in Bristol after having suffered for years from excessive exposure to "plastics, man-made fibers, processed food and gasoline fumes." She lived without venturing forth from her one-woman isolation room, breathed only filtered air, and permitted friends to visit only if they were willing to swear that they had not used toothpaste or deodorant for the previous twenty-four hours, and were not wearing perfume, hair spray, nylon, polyester, or any other synthetic. To gain entry to Rossall's antisynthetic sanctum, guests were obliged to wear pure cotton—regardless of season.

James T. Mclanahan of Woodbury, New Jersey, a 1985 graduate of Dr. Randolph's clinic, had grown used to vomiting a hundred times a day until he confined himself to a room in his house lined with aluminum foil to keep out dust, plastic, and other twentieth-century synthetic debris. He ate only organic food boiled in spring water in a glass pot, and read the newspaper by having it placed inside a sealed metal-and-glass "glove box" similar to those used to handle radioactive materials. He could only watch his tiny TV set for one hour at a time before "the fumes from its plastic parts sickened him" and he was forced to call it a day.

In the shadow of Mount Shasta in the state of Washington, a group of thirty-odd refugees from the Synthetic Century took refuge in a cluster of concrete, glass, and steel houses stripped of all plastic and synthetic substances, including the dread wall-to-wall carpeting. In Potrero, California, in the high desert country east of San Diego, another environmental refugee established what she called the Last Resort, a place where sufferers from EI/MCS could escape from "plastic, polymers and fabric softeners." Some came intending to spend just a couple of weeks but ended up never leaving, for fear of being reexposed to plasticland.

According to Randolph, the major threat posed by plastic was of indoor air pollution caused by the outgassing of plasticizers added during the polymerization process to make plastic soft, flexible, and resilient—all the things that made plastic plastic. "Hard plastics, such as the older Bakelite and Formica, are rarely incriminated as the cause of chronic illness," Randolph maintained. Apparently, the fact that Bakelite was a combination of phenol and the dreaded formaldehyde did not in itself pose a mortal threat to those in proximity to it.

Naugahyde did. "Naugahyde has been particularly troublesome for some patients," Randolph wrote in one of his clinical studies. Legions of EI/MCS sufferers had experienced chronic headaches while trying to "relax" in Naugahyde BarcaLoungers. "The chemically susceptible person," Randolph concluded on a cautionary note, "should try to avoid unnecessary exposure to plastics whenever possible." He recommended eliminating plastic lampshades, because "when the light bulb heats up, the plastic begins to give off odors and fumes that can have a marked effect on mental and physical well-being."

"In the kitchen," Randolph recommended reforming the postwar temple to "safe, unbreakable plastic." "Plastic bowls and dishes should be replaced by ceramic, glass, or wooden ones. Wrap foods in aluminum foil instead of plastic wrap and use glass or metal containers instead of plastic refrigerator ware."

What, no Tupperware, Naugahyde, Saran Wrap—all the marvelous menu of man-made coverings, coatings, and hides that made the Synthetic Century, if not great, at least convenient? It would have been easy enough to dismiss the "chemically sensitive" as neurotics and weirdos. But they could claim—with some justification—that their suffering was in fact taking place for the rest of us. They were serving as "canaries in the coal mine"—a reference to the practice of nineteenth-century coal miners of keeping canaries caged in deep mines to warn them of the possible seepage of odorless poison coal gas into their tunnel. A canary's sudden collapse sent an early-warning signal to the miners that toxic fumes were about to overwhelm the miners themselves. Could the folks at Mount Shasta be sending us a message—falling, predictably, on deaf ears—that plastic was our latter-day golden calf? That if we didn't abandon its promiscuous use and abuse, we were all in imminent danger?

11
Sympathy for the Devil

Courtesy of Fred Kligman

D uring the seventies Peacock Revolution, when newly sensitized "males of the species" were expected to (as the hateful phrase went) "wear the plumage," virgin polyester hit its popular peak. Advances in dyeing synthetics—permitting what one prominent polyester partisan would later derisively term "wild coloration"—produced a dazzling synthetic rainbow of bold and daring shades and textures, adding to the generally birdlike effect.

Years before it became the last word in white-trash aesthetics, double-knit polyester was enthusiastically embraced by the fashion press as the ideal material for the emerging drip-dry, globe-trotting jet set. Even after a red-eye flight in coach, polyester gave fresh meaning to the term "permanent press."

By the latter part of the decade, the white polyester leisure suit worn by John Travolta in *Saturday Night Fever,* with its wide lapels worn over a flared-collar black Qiana nylon shirt—"the most luxurious fabric on earth, imperturbable in its elegance," gushed Du Pont—would sleazily evoke the lower range of the evolving synthetic sensibility, circa 1977.

Polyester's first incarnation, Du Pont's Dacron, prized for its ability to mimic wool, was conservatively tailored to remain as inconspicuous as possible. But the double-knit polyester pants suit was meant to stand out, to boldly announce to the casual observer, "I feel *relaxed.*" Originally designed for traipsing the links or for poolside lounging, the rise of the leisure suit risibly coincided with the ascendancy of the country club, condo culture that took root in the exploding Sun Belt—a Ban-Rol belt, perhaps even equipped with "built-in elastic memory," enabling its wearer to eat up a storm without ever experiencing the nightmare of "waistband pinch." By sheer force of efficiency, stretchable double-knit poly shattered the hierarchy of ancient upper-crust values that had dominated men's fashion since the snuff-sniffing days of Beau Brummell.

In the view of textile historian Samuel Winchester (of the North

Carolina College of Textiles) polyester's aesthetic nadir occurred in the late sixties, when it became a fashion victim of the last decadent days of a declining counterculture. "You could form the garment any way you wanted to," he recalls, appalled at the thought. "You had the drug culture driving all sorts of wild kinds of coloration. All of that merged together in an explosion of leisure suits."

One might argue, if only for the sake of revisionism, that the double-knit polyester leisure suit was more a uniform of Nixon's silent majority than an expression of countercultural rebellion. But Winchester insists that with "the bell-bottoms and double-knits made on the cheap by mass-market producers, the quality went way down. All you had to do was brush up against something and fuzz would develop." Polyester's ultimate downfall, Winchester contends, was the "back-to-nature" movement, "when people started eating natural foods and getting away from chemical-based things." Insofar as the denim-clad denizens of the Woodstock Nation were concerned, polyester was the sartorial equivalent of napalm.

If double-knit polyester represented some sort of synthetic aesthetic nadir, the more positive, heroic, utopian end of the spectrum was occupied by the Jarvik-7 artificial heart, a remarkable plastic utensil developed by a thirty-six-year-old surgical-supply prodigy who patented his first medical invention, a surgical stapler, at seventeen. The first successful implantation of a plastic heart into a human chest cavity was "just like closing Tupperware," recalled surgeon William DeVries, describing the "spiritual" moment in December 1982 when, after working for seven and a half hours to the syncopated beat of Ravel's *Bolero*, he snapped the Jarvik-7 into a set of four Dacron cuffs that had been sewn to the ends of dentist Barney Clark's atria, aorta, and pulmonary artery. After installing a spare plastic left ventricle "we had on the shelf," DeVries watched with satisfaction as Clark's blood pressure steadily rose to a normal 119/75—compared to a feeble 85/40 before surgery.

The remarkable advance in medical technology that gave rise to the Jarvik-7 had been pioneered by Dr. Willem J. Kolff, an acerbic Dutch genius who had managed to scrounge up—in Nazi-occupied Holland—the sheets of cellophane and other synthetic materials required to construct the first kidney dialysis machine. Though Kolff stoically observed sixteen dialysis patients die while hooked to his machine before one finally lived, it remained Kolff's lifelong conviction that "what God can grow, Man can make."

Kolff would continue to tirelessly expound this hubristic thesis after emigrating to the United States in 1950 and founding Kolff Medical—the first surgical supplier of artificial internal body parts—in Salt Lake City, Utah. Kolff became a popular lecturer at the University of Utah School of Medicine. In which guise he inspired twenty-four-year old Robert Jarvik, an American dropout from the University of Bologna medical school, who under Kolff's aegis solved the ultimate puzzle of artificial internal-organ manufacture: the accurate mimicry of human body tissue. Jarvik's groundbreaking research in synthetic "biocompatibility" would permit Barney Clark's body to internalize a plastic Jarvik-7 more readily than a flesh-and-blood human transplant.

Barney Clark was not destined to live very long with that plastic ticker tucked away inside him. But the tidal wave of publicity engendered by Dr. DeVries's ecstatic moment of Tupperware closure would help launch a twenty-year process of public-image rejuvenation comparable in its time only to the metamorphosis of disgraced president Richard Nixon from felon to elder statesman. Like Nixon, plastic would arduously hoist itself out of the depths of disgrace of the Watergate era into a grudging period of public acceptance during the laissez-faire go-go eighties. And like Nixon (who also consistently earned higher approval ratings on the Continent) the beginnings of a new rapprochement between synthetic materials and nature would take root in Europe, where a more sophisticated, less Manichaean grasp of the conflict was adopted by opinion-shaping intellectuals. Plastic's moral rejuvenation owed much of its early momentum to the evolution of a uniquely European postmodern aesthetic, one that placed synthetic materials squarely in the grand tradition of subversive, hierarchy-shattering social forces.

When the distinguished French semiotician Roland Barthes (destined to end his days in an undistinguished fashion, run over by a laundry truck in a Paris street) stopped by a popular exhibition of plastic objects, he was surprised by the length of the line, which stretched around the block and contained mostly parents accompanied by small children. Word, it seemed, had gotten around that it was possible to watch chemicals being turned into useful objects, before your very eyes!

In the ironic long view of the European postmodernists, plastic became something more than just some cheap, middlebrow material to be routinely denigrated by intelligent people; it was a bona-fide social phenomenon that demanded to be understood in terms of the technological evolution of the species. Where Barthes's English and American highbrow counterparts saw nothing but a vast wasteland, Barthes witnessed an act of modern magic. The source of plastic's popularity, Barthes concluded, was that it truly was "the stuff of alchemy." By which he meant that plastic represented the latter-day realization of an ancient, unfulfilled dream—to turn "base" into "noble" metals. The crowds had come "to witness the accomplishment of the magical operation *par excellence,*" Barthes believed— "the transmutation of matter. More than a substance, plastic is the very *idea* of its infinite transformation." To those predisposed to regard it without moral judgment, plastic was an—admittedly often banal—modern miracle.

Even that banality could be seen, when put in the proper perspective, as a virtue: that of humility. Where noble materials—gold, silver, and precious jewels—often engender pride in their wearers and beholders, in Barthes's formulation, plastic was the only "noble" material ever invented that "consents to be prosaic." This was precisely because inherently superior plastic so often disguises itself as something "mundane." In plastic's protean persona—"Its quick-change artistry is absolute: it can become buckets as well as jewels"— Barthes required a transcendent aspect. In its capacity to shatter the ancient hierarchy of values that traditionally separates "noble" from "base" materials, plastic was socially progressive—the first truly democratic material. "And so," Barthes concluded his heartfelt tribute, "the hierarchy of substances is abolished: a single one replaces them all: the whole world can be plasticized, and even life itself, since we are told they are beginning to make plastic aortas."

In postwar Italy, plastic had gained a social position of such prominence unparalleled elsewhere that its mere existence scarcely required moral justification. It was a sign of plastic's prestige, even in the most rarefied academic circles, that plastic rated its own philosopher: the brilliant if often impenetrable Ezio Manzini. An entire generation of Italian designers, artists, and artisans had boldly embraced plastic as the ideal medium with which to build a soaring, sleek new "Olivetti era" on the wreckage of the old order. Pure design and high-minded social engineering would one day merge to form a new, technologically integrated society, a playful plastic paradise—a chemically derived La Dolce Vita.

Manzini celebrated plastic's protean character as part and parcel of a profound paradigm shift in cultural sensibility, from a modern to a postmodern consciousness. Our habitual denigration of surfaces, as opposed to structures, was in Manzini's scheme little more than a sentimental holdover from the heroic days of the Industrial Revolution when "the world was dominated by the looming outlines of locomotives, ocean liners, steel bridges, and great massive constructions."

Plastic's ultimate glory, Manzini maintained, would be realized when the true transcendence of surface—as opposed to substance—was rightfully seen in all its glory, as the reigning symbol of a new society defined no longer by the relationship of concrete objects to each other but rather by "the fluid flow of information." Twenty years before the evolution of the World Wide Web, Manzini was hailing the emergence of the "interface," which he defined as "a semi-permeable osmotic membrane that mediates between two distinct environments." That first semipermeable osmotic membrane was plastic.

In 1995, Edward Rothstein, in an essay published in the *New York Times*, defined an interface as "a surface forming a boundary between two bodies or spaces . . . the membrane that divides one world from another." By then, the interface itself had become the shaper and sculptor of worlds—the chaotic confluence of signs and symbols that comprises the fluid Web. By presenting the first "interface," by elevating surface over structure, plastic had served as the first interface, and a vehicle speeding us toward the cyberspace era.

The inevitable consequence of the triumph of synthetic substances has, according to psychiatrist Robert J. Lifton in his 1994 work *The Protean Self*, helped to define and compel a sweeping psychic shift in human nature. "We are becoming fluid and many-sided," Lifton solemnly contends. "Without quite realizing it, we have been evolving a sense of self appropriate to the restlessness and flux of our time." Lifton has named this new form of consciousness "Protean," in honor of Proteus, "the Greek god of many forms."

Proteanism, in Lifton's view, is a psychic by-product of three distinct social forces: the gradual breakdown of established social and political institutions; the "mass-media revolution" that has provided people with "omni-access" to world events; and the prospect of human extinction, either through nuclear holocaust or ecological catastrophe.

Bioplastic

Plastic has displayed a protean ability to adopt, when necessary, more socially acceptable forms. Since so many people found it utterly alienating, plastic began a gradual and as yet unfinished process of mutation away from the coldly synthetic into a warmer, fuzzier embrace of its traditional enemy, nature.

It's first halting attempt to clean up its act proved more of a setback than an advance. In the early eighties, a number of chemical companies rushed to market—and promoted as "green"—supposedly "biodegradable" plastic bags made mainly out of conventional oil-based plastic, with cornstarch added as filler, which enzymes and other bacteria were supposed to eat in a landfill. But the depressing fact was that once the cornstarch was consumed, 95 percent of the bulk of the bag remained in the landfill—in the form of "plastic dust" that not only would never fully biodegrade but might end up leaching into nearby groundwater.

Even the staunchly conservative *U.S. News & World Report* couldn't help but chortle: "Biodegradables Just Don't Make Degrade." The environmental community was even less charitable, furiously denouncing the first "biodegradables" as a "reprehensible hoax" (The Audubon Society) and a "cynical scam" (Greenpeace).

Following that embarrassing debacle—which included probes launched by a number of states' attorneys general into the implicated company's false and misleading claims—"biodegradability" became something of a joke. But things began looking up on that front when in 1990, researchers at Warner-Lambert in New Jersey announced the development of a biodegradable plastic called Novon made entirely from cornstarch, in which the polymers themselves were made out of starch, not just the filler.

By 1992, Novon was racking up $20 million in annual sales, with further expansion hindered only by cost ($2 a pound, compared to 60 cents for conventional plastics) and by a lack of suitable composting sites able to accept the material. Within a year,

Cargill, the Minnesota-based processor of agricultural products, had developed a rival "bioplastic" made from polylactic acid, a by-product of fermented sugars found in milk, beets, corn, potatoes, and grains.

Though originally discovered in 1833 (and later used in surgical sutures and screws because it breaks down harmlessly in the body) Cargill was able to bring polylactic acid's production costs down from many hundreds of dollars a pound to under $3. Cargill was soon churning out "Eco-Pla" at a rate of 1,200 pounds an hour, while swearing by its ability to be safely recycled, incinerated, or disposed of in a compost heap or landfill, where it was supposed to naturally decompose inside six weeks.

At present, university researchers around the country have been exploring new methods of making plastics without resorting to harmful (and often toxic) petroleum-based chemicals.

At the University of Pittsburgh: Allan Russell and Eric Beckman have been developing a new line of "green" plastics made with so-called supercritical carbon dioxide instead of conventional chemical solvents. Supercritical carbon dioxide is part liquid, part gas, widely touted in the seventies as a possible replacement for petroleum-based solvents.

At the University of North Carolina at Chapel Hill: Joseph DeSimone, a thirty-year-old "wunderkind of chemical engineering" (*New York Times*) has been using supercritical carbon dioxide to generate new kinds of chemical reactions, which are ultimately expected to give rise to a new generation of environmentally benign, solvent-free plastics. Using surfactants—chemicals that form the base for detergents and have long been used to separate molecules in solution—DeSimone has fashioned a nontoxic methyl methacrylate monomer (the primary ingredient in Plexiglas) without having to resort to environmentally unfriendly petroleum by-products.

At Case Western Reserve University (Cleveland, Ohio): Dr. Eric Baer, a professor of micromolecular science, has been studying reindeer antlers—stiffer, tougher, and more shock resistant than bone—to see if he can imitate the rapid cellular process that permits them to grow at the astounding rate of an inch per day. "By studying the composite structure of antlers, we're hoping to create synthetic materials superior to natural antler," Baer explains. Natural materials display "extremely precise control over their structure and composition at every scale, from the molecular level to the level that people can see and feel." A synthetic antler horn, if ever perfected, could make today's tough, durable composite materials resemble eggshells by comparison.

At the Materials Institute at Princeton University: Dr. Ilhand Aksay and Dr. Mehmet Sarikaya have been investigating the deep structure of abalone shell, hoping to synthetically replicate the microthin plates of calcium carbonate that make seashells, on a weight-per-inch basis, one of the hardest, toughest substances on earth—harder and tougher than sheet steel or reinforced concrete. Using a composite material combining boron, carbon, and aluminum, the Princeton researchers have developed a new super-ceramic, far harder than any other man-made ceramic, yet still a weakling compared to abalone.

At the University of Washington in Seattle: Materials scientist Dr. Christopher Viney has learned a secret of silk's extraordinary strength never revealed to Count Chardonnet or his British rival Joseph Wilson Swan. For a microsecond after a spider squirts its water-based solution into the air, the liquid turns into a crystal, instantly snapping the rodlike molecules within it into rigid alignment, "reorienting" it as a polymer engineer might do with a strand of virgin polyester.

Massachusetts Miracle

The fact that polyester no longer serves as the butt of shopping-mall jokes owes much to the collaboration of two visionaries: textiles entrepreneur Aaron Fuerstein and the French-born mountain climber behind Patagonia: Yvonne Chouinard.

In 1961, when Aaron Fuerstein and his father Samuel decided to stay put in western Massachusetts, they bravely bucked a nation-wide trend that had seen the nation's textile industry evacuate the decaying mill towns of the Northeast for ostensibly greener pastures in the low-cost, nonunionized South. Aaron Fuerstein's stubborn decision was based as much on cultural and political will as sound business practices.

"Southern town officials made it clear that we'd be welcome if we didn't bring our liberal Northern ideas with us and silently acceded to discrimination against blacks," he later told the *New*

York Times. "My grandfather came here from Hungary, and I was not about to sell my soul for cheap labor."

The Fuersteins chose to expand Malden Mills, the company founded by Aaron's grandfather in 1906, the year after he arrived from Hungary. In 1956, after taking over an old, abandoned factory complex in Lawrence, a declining mill town thirty-five miles north of Boston where the nation's textile industry first came to life in the 1820s, Malden bet heavily on the market in fake fur—and lost.

By 1981, teetering on the verge of bankruptcy, Aaron Fuerstein, "fighting for [his] life," made another big bet: on a new high-tech polyester product called Polar Fleece, a mountain-climbing and hiking material that mimicked sheepskin at less than half the cost, with greater thermal insulation. Malden Mill's own research-and-development department had perfected the original product, which was bulky, primitive, and double-faced: appropriate outdoor apparel for Conan the Barbarian howling at the gates to the city. But by marketing it mainly through outdoor outfitter Patagonia—which dubbed it "Synchilla"—Fuerstein managed to prevail upon his retailers to identify it by brand name: originally Polar Fleece, which evolved in time into Polartec.

The "tec" in Polartec was key to its success: By manufacturing so-called high-tech fabrics in the heart of a declining mill town, Malden Mills demonstrated that the unionized Northeast was not necessarily a damper to industrial innovation. It also substantially altered the attitude toward synthetics on the part of the one constituency most habitually predisposed to bash them: outdoors men and women. Back-to-nature granola eaters. Tree-huggers. Most of whom, once they tried on a Polartec fleece vest, found their commitment to forswearing petroleum-based products eroding like a badly maintained trail.

By 1992, Malden Mills' arch rival, Wellman, Inc., a *Fortune* 500 synthetic-fiber-manufacturing company based in New Jersey, had emerged with EcoSpun, which turned reviled polyester into a "green" dream by shredding discarded PET bottles into the ultimate in wilderness chic. *Life* captured the giddy role reversal in a headline: "'Cozy, Warm, Yummy, Fleecy'—Is This Any Way to Describe an Empty Soda Bottle?"

To make Ecospun, Wellman gathered thousands upon thousands of PET soda bottles and shredded and crushed them into tiny chips, known as "flake." These, after being washed in a purifying bath, were whitened with bleach if clear, and left alone if tinted green. After being separated by color, the flake was melted and forced through a nozzle to extrude "filaments the width of a human hair." These were then teased and spun into "something totally snug, totally hip—and totally plastic." And totally politically correct.

With the shifting allegiances of the high-minded segment of society *American Demographic* magazine referred to as "visionary greens," the decades-old battle between Nature and Man-Made entered a period of uneasy truce. As for what the magazine calls "hard-core browns"—antienvironmental buyers—they were mainly, the *New York Times* contended, "laughing up their polyester sleeves."

PET Tricks

PET polyester had become one of America's lone bright spots in an otherwise pretty dismal plastic-recycling record. Much to the bottling industry's shame, this relative success has been only due to the passage—over the strenuous protests of beverage and bottling lobbies—of controversial "bottle bills" that made it worthwhile for customers and others to gather up discarded PET bottles for redemption of their nickel-per-container deposits. Nonetheless, the plastics industry points to—*ad nauseam*—the welcome fact that some 30 percent of the PET plastic bottles consumed by Americans in 1993 were ultimately recycled into something useful as a sign that at long last, they are beginning to see the green light.

As well they should—given the increasing preponderance of evidence that plastic may not be the environmentals' number one enemy. Natural substances have problems too. In 1988, NAPCOR (the National Association for Plastic Container Recovery—an industry trade group) commissioned a study from Franklin Associates, an independent environmental research group based in Prairie Village, Kansas, comparing the relative environmental efficiencies of aluminum, glass, and PET.

After repeating its protocol in 1993—while expanding its review to include plastic liquor, fruit-juice, and salad-dressing containers—Franklin's "cradle-to-grave" analysis concluded that "PET soft drink containers have significantly less [environmental] impact than like-size glass bottles by weight, and are equivalent to glass by volume." Glass salad jars were found to generate three times as much atmospheric waste and nearly 70 percent more waterborne waste—by weight—than equivalent plastic containers. The difference was mainly a matter of weight, because heavy glass consumed fossil fuels in transport and cleaning, while aluminum generated substantial waste in its mining and manufacture.

Garbage, Garbage Everywhere

In July 1992, *Atlantic Monthly* managing editor Cullen Murphy and University of Arizona professor William Rathje published *Rubbish! The Archeology of Garbage*. Over a twenty-year period, some 750 scientists and scholars had taken part in the University of Arizona–sponsored Garbage Project, which presided over the painstaking picking over and sorting through of 250,000 pounds of garbage. Approximately fourteen tons of that gross total was excavated from municipal landfills.

In its conclusion, the Garbage Project debunked a number of popularly held "garbage myths' (Murphy and Rathje's phrase) often involving the role of plastic as what the authors referred to as "the Great Satan of Garbage." According to nationwide surveys, a majority of American believed that "fast-food packaging, foam and disposable diapers are major constituents of American garbage." *Rubbish's* authors called this canard "Garbage Myth #1." But after the brown paper bag was unwrapped, the truth (according to Rathje and Murphy) emerged that out of the fourteen tons of garbage from nine municipal landfills analyzed by the Garbage Project over five years, only *a hundred pounds* was composed of fast-food packaging. This amounted, relatively speaking, to a molehill, not a mountain. Fast-food packaging, foam, and disposable diapers, taken together, made up just one-half of 1 percent of landfilled garbage's bulk weight. As for volume—as great a concern as weight at fast-filling landfills—the Garbage Project estimated that fast-food packages took up no more than one-third of 1 percent of the garbage's total volume.

Those figures, the authors hastened to add, failed to include those bulky molded Styrofoam forms used to protect delicate electronic appliances in transit. The vast majority of which, most experts believe, remain stored in the overstuffed attics and basements of the parents of college-age teenagers. But even if those long-suffering grown-ups were suddenly to begin disposing of these foam monsters, the total amount of expanded styrene foam discarded in the United States would probably amount, in Rathje and Murphy's estimation, to "no more than 1 percent of the total volume of landfilled garbage."

Rubbish's Garbage Myth #2 was simply put: "Plastic is a big problem." But the Garbage Project concluded that the volume of all plastics—"foam, film and rigid; toys, utensils and packages"—amounted to around 20 to 24 percent of all garbage. When compacted along with everything else in a landfill, the volume of landfill space taken up by plastic products shrunk to under 16 percent.

Murphy and Rathje's final conclusion was that yes, Virginia, there was a "Great Satan of Garbage." But it wasn't plastic—it was paper. Noting that "a year's worth of copies of the *New York Times* has been estimated to be equivalent in volume to . . . 14,969 flattened Big Mac clamshells," the authors insisted that from "the fifty-year-old newspapers still readable in America's landfills, you could relive the New Deal." Disheartening as the paper problem might be to the literate—and in no way obviated by the false promise of the once widely touted "paperless office"—the Garbage Project concluded that while plastic was certainly no ecological angel, neither was it the devil incarnate.

Fab Gear

In 1958, the mainstay of Du Pont's Teflon team was polymer engineer Wilbert L. Gore. Within the company, Gore had fought long and hard to exploit Teflon's potential to shine as a consumer as well as an industrial product. But though nonstick Teflon-coated omelette pans made their debut in France in early 1961 (and at Macy's in New York a year later) Du Pont officials remained uneasy about Teflon's consumer potential. Most disturbingly, at high temperatures, Teflon displayed a tendency to "outgas" fumes that, though not strictly toxic, had been implicated in the production of "flulike" symptoms.

Despite such drawbacks, Gore remained a staunch advocate for Teflon. In the late fifties, after identifying a market opportunity for specialized electronic cable coated with Teflon, he was bitterly disappointed when Du Pont's executive committee declined to encourage him to exploit it.

So Gore quit—and in partnership with his wife Vieve, set up shop in the basement of their home to make PTFE (polytetrafluoroethylene) coated cables. The company prospered, in a modest way until 1969, when Bill Gore's son Bob, a former polymer-engineering star student (who had since risen to president of the small family company) discovered that by yanking on a rod of Teflon—instead of gently and gradually pulling it—he could stretch a piece of Teflon into a porous, superstrong filament.

Expanded Teflon turned out to be a superb transmitter of computer data, an excellent biologically inert material for surgical supplies, and last but not least, when bonded to a synthetic outer layer, the first truly "poromeric" material—a polymer that actually breathed. Du Pont's Corfam dream come true.

Waterproof clothing that breathed like cloth had been the goal of every plastics pioneer from the early days of Charles Macintosh and his rubberized-wool Macks, Thomas Hancock and his rain-soaked carriage passengers, Charles Goodyear and his sulfurized rubber coats, Daniel Spill with his collodion-coated cotton, and Alexander Parkes with his Parkesine-coated canvas.

But even with the evolution of vinyl (World War II's best attempt at waterproof rain gear), plastic's problems had seemed insurmountably similar to those of rubber: It formed an effective vapor barrier, but also an oxygen barrier. With Gore-Tex, Gore had engineered a "hydrophobic" (water-resistant) membrane that also managed to remain "air-permeable." As the friendly folks at Gore like to say: "It keeps water and wind out while allowing perspiration to escape."

The Wicking Witch of the West

"Wicking" is high-tech talk for "pulling moisture away from the body while still insulating it." Put even more simply, "wicking" permits you to perspire while staying dry, because the moisture is drawn away from the surface of the skin to the outermost layer of fabric.

To test synthetics against natural materials while on a Minnesota vacation, Neal Karlen (reporting from lakeside for the *New York Times)* ventured out on a cross-country ski field one windy fourteen-degree day wearing a cotton flannel union suit, cotton sweatpants, a hooded cotton sweatshirt, and a woolen sweater. "Within minutes," he would later recall, "I was sweating like a sumo wrestler." The problem? He was unwicked. Within half an hour, he was nearly frozen solid.

The next day, Karlen went ice-skating dressed from head to toe in the latest high-performance synthetics: a tasseled wool hat from Dale of Norway ($34) lined with a polyester fabric from Gore-Tex called Gore-Wind Stopper, supposed to "warm the ears while avoiding wool's scratchiness." His neck was encased in a Patagonia recycled-PET Synchilla neck-gaiter ($15), while next to his skin he sported a pair of Early Winter Prolite 5000 polypropylene long johns. Sweat wicked off quickly, keeping him dry as a bone no matter how hard he skated, or how much he perspired.

When it began to snow thickly, he supplemented his outfit with a North Face heavy fleece Denali insulating jacket ($160), topped by Gore-Tex-lined shell by Marmot ($200). A nifty pair of Outdoor Research's two-fingered Gore-Tex gloves ($65) completed the ensemble. Total cost (not including ice-skating): $475.

When Karlen stomped out to some friends' ice-fishing hut in the middle of a frozen lake decked out in his high-priced getup, "the tough old guys" standing for hours on the ice "wearing woolens" estimated that for the price of all that "wicking" and "breathing" he could have picked up a new aluminum fish shack, a portable ice auger, and "enough minnows to last until the next millennium." But Karlen was warm and dry. The score: Synthetic materials 1, Nature 0.

Back to Basics

Plastic has lately mutated into less of an outright offense to and assault on nature by quietly returning to its century-old roots as a protector of endangered species. In the 1860s, with the population of elephants from the island of Ceylon (now Sri Lanka) mostly wiped out (making billiard-ball ivory "dreadfully dear"), John Wesley Hyatt developed celluloid, partly as a way of saving the endangered African elephant from extinction—and to keep Americans playing pool.

Today, with the African black rhinoceros destined to die out any day now, Robert Molloy, a professor of polymer engineering at the University of Massachusetts at Lowell, has teamed up with John Linehan (curator of the Franklin Park and Stone zoos in the same state) to develop a plastic black rhino horn.

With only 2,800 black rhinos left in Africa (down from about 70,000 in 1970) Molloy and Linehan remain gravely concerned that there may not be much time to waste. They hope to enlist the help of game wardens to tranquilize the rare rhinos, remove their horns, and implant the composite plastic horn over the real rhino horn stump like a dental crown.

Namibia has already nixed the plastic plan because it claims to be satisfied with its own program of removing the horns and not replacing them. But "rhinos need horns in order to be socially compatible," Molloy maintains. "They use them to make critical signals defining dominance hierarchies." As a deterrent to poaching, the plastic rhino horn would be colored orange, to alert would-be attackers that it is a (costly) fake. The estimated cost of between $1,500 and $2,000 per horn pales next to the worth of the ivory original: $100,000.

In the mid-eighties, the importation of elephant ivory was banned by most civilized countries because of the ceaseless slaughter of African and Asian elephants. Among the worst sufferers were concert pianists, who much as they might desire elephants to remain in the wild, nevertheless had become attached to the apparently irreplaceable feel and action of genuine ivory keys.

Looking to the future, the Steinway piano company recently commissioned a team of polymer engineers from Rensselaer Polytechnic Institute in Troy, New York—all experts in "the study of the sliding or movement of surfaces as they relate to bearing and seals"—to come up with a more ivorylike plastic piano key. After examining ivory under an electron microscope, they measured the surface roughness of the real ivory keys, looking closely at the height of the peaks and valleys to see how they were oriented to each other. Much to their surprise, studies showed that the orientation was entirely random.

The ivory pores that subtly absorbed fingertip moisture turned out to be randomly scattered across the surface as well. So the polymer engineers simply took eleven real-ivory keys donated by Steinway and made precise molds from them, tested a wide variety of possible plastics to "evaluate the friction against the fingers," and in the end settled on a polyester mixed with water-soluble wax beads. After the resin sets, the wax beads are washed out, leaving randomly patterned pores.

Plastic Nostalgia

For the nineties, Tupperware (now a division of a conglomerate called Premark International) has gone multicultural. Getting in touch with the urban market, the *New York Times* recently reported, requires "a sales force that can hip-hop, samba, polka and kvetch. . . . Whether they eat arroz con pollo or challah or moo goo gai pan," Tupperware has a thirty-ounce Thirstquake tumbler or Snackalizer tray to keep their favorite foods fresh and fancy.

Tupperware has been redesigned for the Millennium by sixty-year-old Brooklyn-born Pratt graduate Morrison Cousins, who is trying to do for Tupperware in the nineties what Donald Deskey did for desktop staplers in the thirties: bring them up to date. "His sleek products bring high Eurostyling" to the Tupperware table, says curator Kathryn Hiesinger of the Philadelphia Museum of Art, which has recently added a selection of Cousins's new Tupperware tableware to its collection.

Yet surely as plastic has given, plastic has taken away. Author Mary Blume, in a bittersweet account of the rise and fall of France's fabled Côte d'Azur, reports that at Menton's annual *Fête du Citron*, plastic lemons outnumber real ones. The world's beaches—according to a recent editorial in the *New York Times*—suffer from a growing shortage of sea-glass: those colorful matte-surfaced shards snapped off old smashed bottles that children and adults have collected for years, often to gather dust on the pine shelves of beach houses. These days, if you want sea-glass, you have to buy it in a ten-ounce bag, which mixes artificially ground sea-glass with a smattering of shells and coral to make Beach Debris (retail cost $20), which comes complete with a scent designed to evoke the sea. The real culprit is not the merchant but an overabundance of plastic bottles that never break, and when they do wash ashore, produce disgust, not delight.

And yet, and yet: "Jellies" are back! Those semitransparent, colorful buckle-on cheap plastic shoes that once, if dared to be worn in America, marked their wearer as a confirmed Euro-poseur. Courrèges is back, though not by name, but by the sincerest form of flattery: a fashionable flurry of white vinyl coats that nostalgically evoke his space-age vision—though "with a look," the *Times* reports, "much less severe than Courrèges."

The rampant popularity of recycled plastic fleece prompts one women's magazine to worry: "Is there any risk of the real fake thing—'classic polyester'—getting elbowed out of the market completely?" Only to be reassured by Malden Mills, which stoutly insists: "As long as there's oil, you'll still be able to get virgin polyester."

In *The New Yorker,* author Susan Orlean bashes Polar Fleece for having "no historical, cultural, or physical association with a place, a season, a society, or any living thing. It is the first existential fabric—eminently useful, meaningless, dissociated and weird."

Yet polyester is again riding high, now in finer form, as "microfiber"—accepted on the best runways and admitted into the chicest salons. If people know that "microfiber" means "polyester," they tend to politely forget it. As one pundit put it, "Just when you thought it was safe to go back into the stores, microfiber appeared like a white knight to rescue disco duds from distress."

The gutters of Paris are still swept by street sweepers carrying twig brooms. But the "twigs" are green plastic. American Express and Sotheby's have agreed to cancel an arrangement permitting buyers of art to pay for paintings "with plastic" after Los Angeles collector Eli Broad picked up (appropriately) Roy Lichtenstein's *I . . . I'm Sorry* for $2.5 million, entitling him to 2.5 million miles on any airline that participates in AmEx's frequent-flyer program.

To spread the scent of spring, scientists at International Flavor and Fragrance have developed "polyiffs"—microscopic polyester capsules that release the aroma of fresh flowers "at a measured pace to simulate the smell of a spring garden." Virtual smell-o-vision!

The plastic-shelled '53 Corvette has been reincarnated as the GM Ultralite, a four-passenger "concept car" that weighs 1,400 instead of 3,000 pounds because of its carbon-fiber composite body—"The same stuff they make tennis rackets out of," notes a GM spokesman.

At the Museum of Modern Art in New York, design curator Paola Antonelli, a former editor with the Italian magazine *Domus,* has put together a show called "Mutant Materials," the high point of which was a rubbery, goopy New Wave Silly Putty–esque stuff intended for exercising the forearm, which comes in a rainbow of "yummy neon colors."

The show focused a good deal of welcome and belated aesthetic attention on such high-tech hybrids as "viscoelastic-polymer" and "thermoplastic vulcanite." The room was dominated by groovy, New Wave sports equipment, where the true material innovations are happening: ski poles and kayaks, brilliant blue blobs of bicycle seats, wryly described by *New York Times* design critic Hubert Muschamp as "uniformly frisky and sexy as an episode of *Baywatch.*"

In an accompanying catalog essay, Ms. Antonelli cites the usual suspects of postmodern deconstructivist criticism, with a tad of chaos theory thrown in, to reinforce her assertion that "revolutions in science, technology and philosophy are historically accompanied by deep, unconscious shifts in the culture at large."

In short: We mold plastic; plastic molds us.

Plastic Pilgrimage

It may be a far cry from the Museum of Modern Art to the National Plastics Center and Museum in Leominster, Massachusetts, but I couldn't conceive of not stopping by the home of the John Wesley Hyatt Society to pay my respects to the man who got the plastic billiard ball rolling. Leominster is still something of a center for the U.S. plastics industry; the home among other thriving enterprises of Union Products, where Don Featherstone, creator of the original plastic pink flamingo, still sculpts and sells lawn art by the ton—though now in the guise of an internationally known, world-famous seminal artist: He signs his pieces.

In the reception area of the world's only Museum of Unnatural History, a Don Featherstone Original Pink Flamingo greets visitors with an upturned beak and winsome smile. Hoping to find the plastics museum a little poky, a little rinky-dink, I was due for disappointment. Carved out of a handsome old brick schoolhouse, it is a thoroughly modern, sophisticated, vest-pocket facility plainly and seriously dedicated to celebrating the unsung role that plastics have played in modern life. The main attraction, and centerpiece of the Environmental Gallery, is the Micro Molder, a massive Rube Goldberg–esque contraption that takes old polyethylene milk bottles, chops them up, and turns them into useful key chains.

On the Saturday afternoon of my visit, a long line of towheaded Cub Scouts, each wearing a bright orange Tiger Cub T-shirt and clasping an HDPE milk jug in his hand, looked on goggle-eyed as education director Barb Bennet, wearing plastic goggles, stood astride the great roaring-and-thumping machine gamely shouting above the Victorian din:

"First you're going to open the hopper. Then you're going to throw in the milk jug."

As the shambling first-graders moved one by one to the top of the line and meekly complied with these directives from above, I spied neatly framed between two clanking gears a sign on the wall:

FROM POP BOTTLE TO POPLIN: POLYESTER
BECOMES ENVIRONMENTALLY COMPATIBLE.

Yessiree Barb

As Barb energetically tugged on her long lever, she patiently explained at the top of her lungs that the white snowy polyester flake being hosed out of the hopper was being heated to five hundred degrees, and would soon melt into liquid form. As it did just that, the molten mass exuded a pungent odor I distinctly recalled from my own long distant schooldays. Vacuform! A plasto-Proustian moment.

"You get seven key chains from just one plastic milk bottle," Barb shouted encouragingly.

"It's melting!" Barb cried out, joyfully pointing at the sharp smell's source.

"Do you guys know what the plastic is now?"

The kids hesitate, mumbling under their breaths.

"It starts with a *P*," Barb playfully prompts.

In unison, the well-coached voices shrilly shout: *"PLIABLE!"*

Though not yet carved on the tabula rasa walls of the National Plastic Museum, perhaps the following ancient verses should be:

Human beings are born soft and supple
At their death they are stiff and hard
The grasses and trees are born tender and pliant
At their death they are withered and dry.
Therefore the stiff and inflexible
is a disciple of death
The gentle and yielding
is a disciple of life.
An army that is inflexible never wins a battle.
A tree that is unbending easily snaps.
The hard and rigid will be broken.
The supple and yielding will prevail.

—Lao-tzu
Tao-te ching (as quoted in a Shell Chemical ad
in *Modern Plastics*, July 1982)

Index